ACT Math Personal Tutor

David Ebner, Ph.D.

NOVA PRESS

ACT is a registered trademark of ACT, Inc., which was not involved in the production of, and does not endorse, this book.

ACT Math Personal Tutor

The Author, David Ebner, has a Ph.D. from NYU. He has taught for many years at the high school and college levels.

His previous publications include:

- Math Word Problems the Easy Way, Barron's
- How To Prepare for the SSAT/ISEE, Barron's
- How To Prepare for the COOP/HSPT/TACHS, Barron's
- SSAT/ISEE, Barron's
- COOP/HSPT/TACHS, Barron's
- Elementary Algebra, United Federation of Teachers

Copyright © 2018 by David Ebner. No part of this publication may be reproduced, stored or distributed in any form without prior written permission from the author.

ISBN 10: 1-944595-79-1
ISBN 13: 978-1-944595-79-1

Published by Nova Press

P.O. Box 692023
West Hollywood, CA USA 90069
1-310-275-3513
info@novapress.net
www.novapress.net

ABOUT THIS BOOK

Why this ACT math review book is different from other review books?

- Formulas and definitions are listed right at the start of the book because they form the backbone of the ACT. Make certain that you understand them and their applications before you begin working your way through the sample tests.

- Cartoons with suggestions for solving the problems are included on the first two tests. Five full-length exams are included.

- Lots of white space makes the material much easier to read and comprehend.

- Detailed and easy-to-understand solutions are provided for each exam.

- Questions are repeated on the answer pages so that you don't have to flip through the pages to get back to the questions.

All of the topics in the ACT are covered, and a table for converting raw math scores into scaled scores is included at the end of the book. A second table converts scaled math scores to percentiles. To get an idea of your own strength in mathematics, use this key to score each test.

Hints for Scoring High on the Math Section of the ACT

1. There are no penalties for guessing, so fill in all the bubbles.
2. There are 60 questions and 60 minutes allocated, so try to take one minute per question.
3. The raw scores are converted to scaled scores and then percentiles. You can check your scores and percentiles at the back of this book.
4. Use a digital watch to check your time.
5. If you're not sure how to solve the question, plug in the answer.
6. Eliminate the obviously incorrect choices first. Then go on to solve the problem.

Format of the Math Section

The math section contains 60 multiple-choice questions and is 60 minutes long. The test measures mathematical skills typically obtained through grade 11 or 12, that is, through Trigonometry.

The questions are listed roughly in ascending order of difficulty. The section typically begins with Pre-Algebra questions, and progresses to Elementary Algebra, then to

Intermediate Algebra, and then finally to Trigonometry. But there can be considerable overlap in these categories.

Section	Type	Time
Math	60 Multiple-choice Questions	60 minutes

Here are the approximate percentages of the content categories on the math section:

1) Pre-Algebra/Elementary Algebra (40%)
2) Intermediate Algebra/Coordinate Geometry (30%)
3) Trigonometry/Plane Geometry (23%)

The math section is always the second section of the test.

Scoring the ACT

There are four scores recorded:

1) Total score (based on all 60 questions)
2) Elementary Algebra (based on 24 questions)
3) Intermediate Algebra (based on 18 questions)
4) Trigonometry (based on 18 questions)

The total score is reported on a scale from 1 to 36, with 36 being the highest score possible. And the sub-scores are reported on a scale from 1 to 18, with 18 being the highest score possible. The average total score is about 21. Do not become discouraged if you "blow" a few questions. The average ACT student misses more than half of the questions on the math section.

Odds and Ends

- You can take the ACT up to 12 times, and you get to choose which exam to submit to the college of your choice.
- Most students take the exam for the first time in their junior year of high school and then again in their senior year.
- You are allowed a calculator, but not one that offers a computer algebra system (e.g., TI-89). Check whether your calculator is permitted.
- Wherever possible, use your graphing calculator.
- For further questions, contact www.act.org.

Good luck on your test!

David Ebner

CONTENTS

About This Book	3
Formulas and Definitions	7
Sequences	7
To Find any Term in an Arithmetic Sequence	7
To Find the Sum of an Arithmetic Sequence	7
To Find any Term in a Geometric Sequence	7
To Find the Sum of a Geometric Sequence	7
The Quadratic Formula	8
Parabolas	8
Areas	9
Area of a Triangle	9
Area of a Circle	9
Area of a Parallelogram	10
Area of a Trapezoid	10
Area of a Regular Polygon	10
Volumes	10
Volume of a Cylinder	10
Volume of a Rectangular Solid	10
Volume of a Cube	10
Variation	10
Direct Variation	10
Inverse Variation	10
Joint Variation	10
Trigonometry	11
Definition of Trigonometric Ratios	11
Graphing Trigonometric Functions	11
Trigonometric Identities	11
Pythagorean Identity	11
Sum and Difference Formulas for Angles	11
Exponents	12
Logarithms	12
Equation of a Circle	13
Probabilities	13
Matrices	14
Adding Matrices	14
Subtracting Matrices	14
Multiplying Matrices	14
Functions	15
Introduction to Most Calculators	15

Test 1	**17**
Answer Key	85
Test 2	**87**
Answer Key	163
Test 3	**165**
Answers	177
Answer Key	214
Test 4	**215**
Answers	228
Answer Key	269
Test 5	**271**
Answers	287
Answer Key	327
Converting Raw Scores to Scaled Scores	328
Converting Scaled Scores to Percentiles	329

FORMULAS AND DEFINITIONS

Because it can be rather dull to spend a lot of time reviewing basic math before tackling full-fledged ACT problems, this chapter presents just some foundational knowledge of mathematics. Then, in later sections, each problem introduces a new math concept. Through this method, all the math you need for the ACT will be reviewed.

Sequences

To Find any Term in an Arithmetic Sequence, Use the Following Formula:

$$a_n = a_1 + (n-1)d$$

where a_n = the nth term, a_1 = the first term, n = the number of terms, and d = the difference between any two successive terms.

To Find the Sum of an Arithmetic Sequence, Use the Following Formula:

$$S_n = \frac{n}{2}(a_1 + a_2)$$

where S_n = the sum of n terms, n = the number of terms, a_1 = the first term and a_n = the nth term. What we're actually doing here is getting the average of the first and last terms and multiplying the average by the number of terms in the sequence:

$$S_n = (\text{number of terms})(\text{average}) = (n)\left(\frac{a_1 + a_2}{2}\right)$$

To Find any Term in a Geometric Sequence, Use the Following Formula:

$$a_n = a_1 r^{n-1}$$

where a_n = the nth term, n = the number of terms, a_1 = the first term, and r = the common ratio between any two successive terms.

To Find the Sum of a Geometric Sequence, Use the Following Formula:

$$S_n = a\left[\frac{1-r^n}{1-r}\right]$$

where S_n = the sum of n terms, a = the first term, r = the common ratio, and n = the number of terms in the sequence.

The Quadratic Formula

We can use the following formula to solve equations of the form $ax^2 + bx + c = 0$, where a, b and c are real numbers and $a \neq 0$:

$$x = \frac{-b \pm \sqrt{b^2 - 4ac}}{2a}$$

Parabolas

$$y = ax^2 + bx + c$$

$y = x^2$ $\qquad\qquad$ $y = -x^2$ $\qquad\qquad$ $y = x^2 + 3$

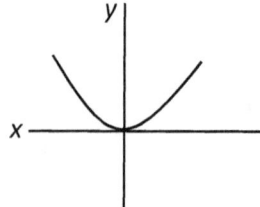 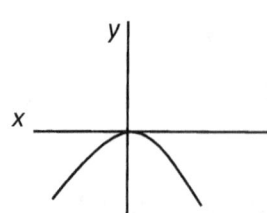 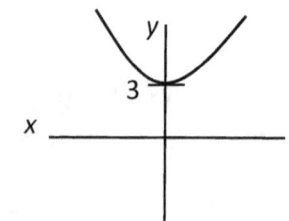

$y = 2x^2$ $\qquad\qquad$ $y = 2x^2 + 2x$ $\qquad\qquad$ $y = 2x^2 + 2x + 3$

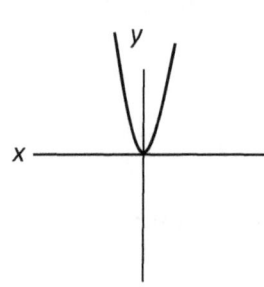 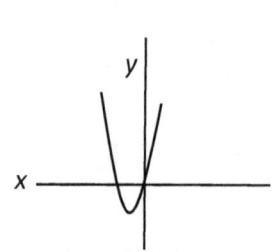 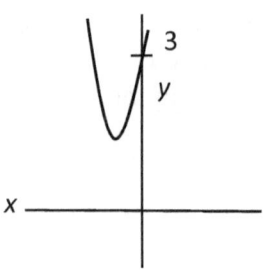

$y = (x-1)^2$ $\qquad\qquad\qquad$ $y = (x-1)^2 + 3$

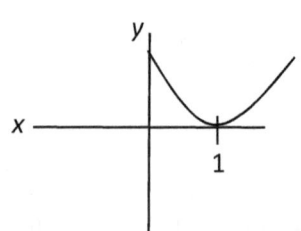 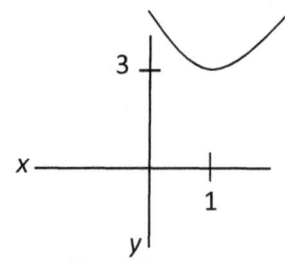

Formulas and Definitions

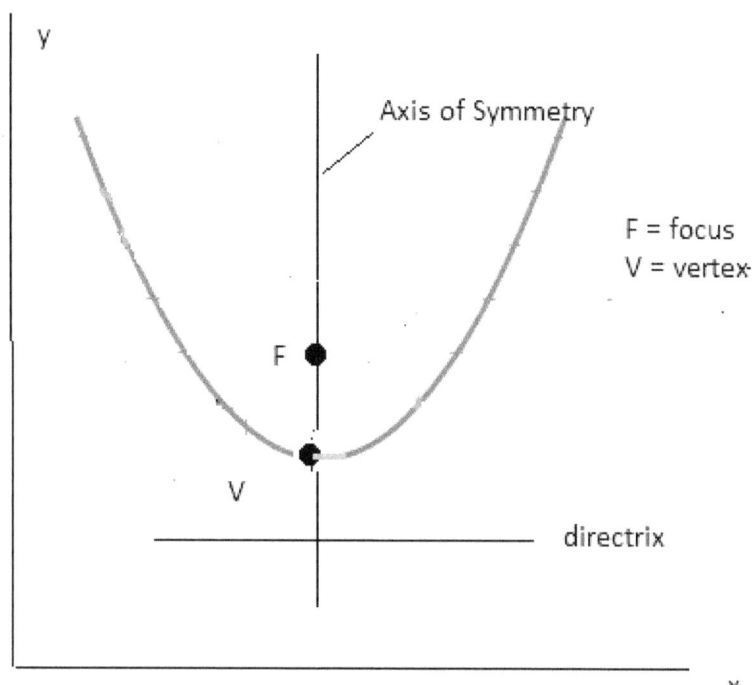

A standard form for a parabolic equation is $y = ax^2 + bx + c$. In this case, the x-coordinate of the vertex is $x = \frac{-b}{2a}$, and it is also the equation of the axis of symmetry.

In the quadratic equation $y = ax^2 + bx + c$, when $a < 0$, the parabola faces downward.

Areas

Area of a Triangle
1. $A = \frac{1}{2} bh$, where A = area, b = base, h = height

2. $A = \sqrt{s(s - a)(s - b)(s - c)}$, where s = semi-perimeter and a, b and c are the sides of the triangle.

3. $A = \frac{1}{2} ab \sin C$, where a and b are two sides of the triangle and C is the included angle.

Area of a Circle
$A = \pi r^2$, where r = radius

Area of a Parallelogram
$A = bh$, where b = base and h = height

Area of a Trapezoid
$A = \frac{1}{2} h(a + b)$, where h = height, a = one base and b = the other base

Area of a Regular Polygon
$A = \frac{1}{2} ap$, where a = apothem and p = perimeter

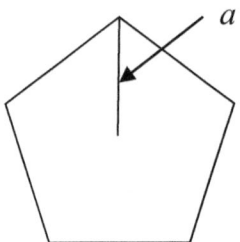

Volumes

Volume of a Cylinder
$V = \pi r^2 h$, where r = radius, and h = height

Volume of a Rectangular Solid
$V = lwh$, where l = length, w = width and h = height

Volume of a Cube
$V = e^3$, where e = length of the edge (or side) of the cube

Variation

Direct Variation
$y = kx$

In this equation, y varies as the constant factor of proportionality k.

Inverse Variation
$xy = k$

In this equation, x and y vary in opposite directions. The constant k is the constant of proportionality.

Joint Variation
$y = kxz$

In this equation, y varies as the joint product of x and z. The constant k is the constant of proportionality.

Trigonometry

Definition of Trigonometric Ratios

SOHCAHTOA is a mnemonic (memory devise) for these ratios of the sides of a triangle.

$\sin \theta = \dfrac{opposite}{hypotenuse}$ $\qquad$ $\cot \theta = \dfrac{adjacent}{opposie}$

$\cos \theta = \dfrac{adjacent}{hypotenuse}$ $\qquad$ $\csc \theta = \dfrac{hypotenuse}{opposie}$

$\tan \theta = \dfrac{opposite}{adjacent}$ $\qquad$ $\sec \theta = \dfrac{hypotneuse}{adjacent}$

Graphing Trigonometric Functions

$$f(x) = a \cos bx + c, \text{ where } x \text{ is in radians}$$

1. $|a|$ is the amplitude
2. $\dfrac{2\pi}{b}$ ($b > 0$) is the period
3. $+c$ indicates a vertical shift upwards
4. $-c$ indicates a vertical shift down

The same description holds for the sine function.

Trigonometric Identities

$\csc \theta = \dfrac{1}{\sin\theta}$ $\qquad$ $\tan \theta = \dfrac{\sin\theta}{\cos\theta}$

$\sec \theta = \dfrac{1}{\cos\theta}$ $\qquad$ $\cot \theta = \dfrac{\cos\theta}{\sin\theta}$

Pythagorean Identity

$$\sin^2\theta + \cos^2\theta = 1$$

Sum and Difference Formulas for Angles

$\cos(A + B) = \cos A \cos B - \sin A \sin B$

$\sin(A + B) = \sin A \cos B + \cos A \sin B$

$\cos(A - B) = \cos A \cos B + \sin A \sin B$

$\sin(A - B) = \sin A \cos B - \cos A \sin B$

$\tan(A + B) = (\tan A + \tan B) / (1 - \tan A \tan B)$
$\tan(A - B) = (\tan A - \tan B) / (1 + \tan A \tan B)$

Exponents

$x^a x^b = x^{a+b}$

$(x^a)^b = x^{ab}$

$\dfrac{x^a}{x^b} = x^{a-b}$

$x^{\frac{1}{n}} = \sqrt[n]{x}$ This definition coverts fractional notation to radical notation. For example, $x^{\frac{1}{3}} = \sqrt[3]{x}$.

$x^0 = 1$ This is a definition, not a property. That is, x^0 is defined to be equal to 1.

Logarithms

Definition of Logarithm:

$$\log_b x = y \Leftrightarrow b^y = x$$

The double arrow means "if and only if."

Note that b is called the base, x is the number whose log is being calculated, and the result y is the log. Both x and b must be positive.

Students often find logs mysterious. Even after studying logs for a while, students may still ask, "what is a log?" It is nothing more than the definition above! Notice that on the right side of the definition that y (the log) is an exponent: it is the power to which you raise the base to get the number you are taking the log of. If this seems contrived, you're right. A log is a pure contrivance.

Let's use this definition (rule) to calculate some logs:

1) $\log_2 8 = y$

   Here, the base b is 2, x is 8, and y is itself. Plugging this information into the right side of the definition yields

   $$2^y = 8$$

   Clearly, the solution to this equation is 3: $2^3 = 2 \cdot 2 \cdot 2 = 8$.

2) $\log_3 27$

Here, we are being asked to find to what power do you raise 3 to get 27. Though y was not written in the problem, we will use it in the solution. Here, the base b is 3, and x is 27. Plugging this information into the right side of the definition yields

$$3^y = 27$$

Clearly, the solution to this equation is 3: $3^3 = 3 \cdot 3 \cdot 3 = 27$.

$\log x = \log_{10} x$ (the common log)

$e = 2.71828...$

$\ln x = \log_e x$ (that is, log base e is called the natural log)

$\log_b a = \dfrac{\log_c a}{\log_c b}$ this is the change of base formula, where c is any positive number

$\log_b \dfrac{c}{d} = \log_b c - \log_b d$

$\log_b c^d = d \log_b c$ the "leap frog" rule (with this silly name, you are unlikely to forget the rule)

Equation of a Circle

When the center is at the origin, (0, 0) and (x, y) is a point on the circumference:

$$x^2 + y^2 = r^2$$

When the center is at (h, k) and (x, y) is a point on the circumference:

$$(x - h)^2 + (y - k)^2 = r^2$$

Probabilities

Probability of Mutually Exclusive Events Occurring

$$P(A \text{ and } B) = P(A) \cdot P(B)$$
$$P(A \text{ or } B) = P(A) + P(B)$$

Probability of Non-Mutually Exclusive Events Occurring

$$P(A \text{ or } B) = P(A) + P(B) - P(A \text{ and } B)$$

Matrices

Matrices afford an efficient way of solving systems of equations. We won't study this method, however, because it is unlikely on the test that you will be asked to solve a system of equations by using matrices. Instead, you will be given problems that require an understanding of the properties of matrices.

A *matrix* is just an array of numbers. Some examples are

$$\begin{bmatrix} 1 & 2 & 3 \end{bmatrix} \qquad \begin{bmatrix} 1 & 2 \\ 3 & 4 \end{bmatrix} \qquad \begin{bmatrix} a & b & c \\ d & e & f \\ g & h & i \end{bmatrix}$$

In general, an $m \times n$ matrix (m rows and n columns) is

$$\begin{bmatrix} a_{11} & a_{12} & a_{13} & \cdots & a_{1n} \\ a_{21} & a_{22} & a_{23} & \cdots & a_{2n} \\ a_{31} & a_{32} & a_{33} & \cdots & a_{3n} \\ \vdots & \vdots & \vdots & \ddots & \vdots \\ a_{m1} & a_{m2} & a_{m3} & \cdots & a_{mn} \end{bmatrix}$$

Adding Matrices

To add two matrices, just add their corresponding elements:

$$\begin{bmatrix} a & b \\ c & d \end{bmatrix} + \begin{bmatrix} e & f \\ g & h \end{bmatrix} = \begin{bmatrix} a+e & b+f \\ c+g & d+h \end{bmatrix}$$

Subtracting Matrices

To subtract two matrices, just subtract their corresponding elements:

$$\begin{bmatrix} a & b \\ c & d \end{bmatrix} - \begin{bmatrix} e & f \\ g & h \end{bmatrix} = \begin{bmatrix} a-e & b-f \\ c-g & d-h \end{bmatrix}$$

Multiplying Matrices

To multiply two matrices, multiply the elements in each row of the first matrix by the corresponding elements in each column of the second matrix. The number of rows must equal the number of columns.

$$\begin{bmatrix} a & b \\ c & d \end{bmatrix} \begin{bmatrix} e & f \\ g & h \end{bmatrix} = \begin{bmatrix} ae+bg & af+bh \\ ce+dg & cf+dh \end{bmatrix}$$

Functions

A function is a relationship in which each element in the domain of the function is matched with one and only one element in the range of the function.

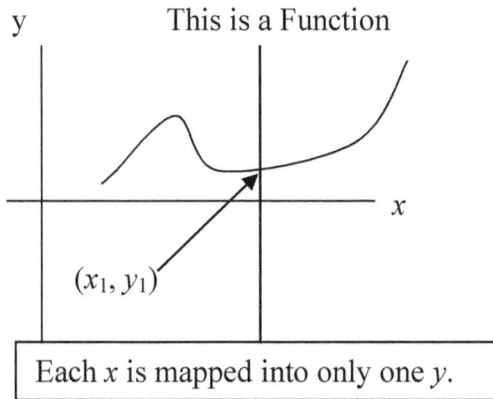

Each x is mapped into only one y.

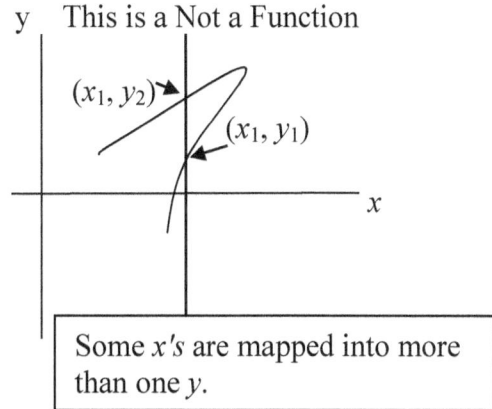

Some x's are mapped into more than one y.

Introduction to Most Calculators

Usually, each key on a calculator has three functions.

For example, *log* has three functions:

sample key

| 10^xN |
| log |

1) If we just press *log* and enter a number, say 100, and then close the parentheses and press *enter*, we get 2. When we use *log*, we mean base 10, so $\log_{10} 100 = 2$ (since $10^2 = 100$).

2) When we press 2nd + *log* and enter a number, say 2, we get 100. This means 10^2, which is 100.

3) To get to the third function, press *alpha* and then *log*. We get the variable N. We can use the alpha mode to enter words or to enter titles to columns in data lists.

Raising a Number to a Power

Let's take 5 raised to the power of 3 as an example: we enter 5^3 and then press Enter, which returns 125 on the calculator.

Inverse Functions

Let's take the reciprocal of 4: 4 + (the button) x^{-1} and then press Enter, which returns .25 on the calculator.

Graphing algebraic functions on a calculator

Go to *y =*. Enter the function.
Enter *Graph*.

For example, to graph the function 2^x, enter *2^x* and then press *Graph*. The calculator should then display the following graph:

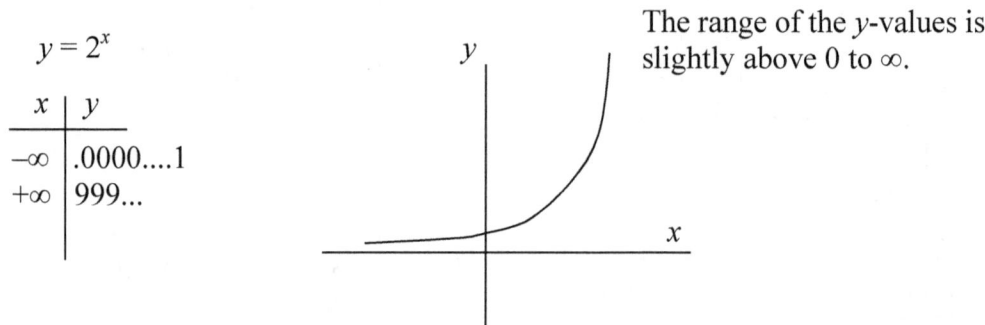

$y = 2^x$

x	y
$-\infty$	.0000....1
$+\infty$	999...

The range of the *y*-values is slightly above 0 to ∞.

Trigonometric Functions

Go to the calculator function *Mode* and change the setting (radians, grad, etc.) to degrees (if the calculator is not already in degree mode) and then perform the following calculation:

Trig Function	Inverse Function
input: *sin* (30°)	input: 2nd, *sin* (.5)
the answer should be .5	the answer should be 30°

Graphing Trigonometric Functions.

1) $y = 2\sin x$ (the mode setting on the calculator is radians)
   Graph

2) $y = 2x^2 - 6$
   Graph

Test 1

1. What is the quotient of the least common multiple of 18 and 24 and their greatest common divisor?

(A) 18 (B) 15 (C) 58 (D) 12 (E) 78

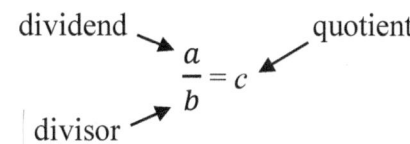

The least common multiple of two integers is the smallest multiple which both numbers have in common. The greatest common divisor of both numbers is the largest divisor in common.

dividend → $\dfrac{a}{b} = c$ ← quotient
divisor ↗

2. The Syracuse Panthers baseball team won 54 games and lost 16 games. If they want to get an 80% winning average, how many more games do they have to win without any losses?

(A) 8 (B) 10 (C) 12 (D) 14 E) 9

Develop the ratio of the number of winning games to the total number of games and set it equal to 80%.

17

ACT Math Personal Tutor

Solution:

1. What is the quotient of the least common multiple of 18 and 24 and their greatest common divisor?

(A) 18   (B) 15   (C) 58   (D) 12   (E) 78

$$\left.\begin{array}{r}18 \times 4 = 72 \\ 24 \times 3 = 72\end{array}\right\} \text{72 is the least common multiple}$$

Note: Often a quick way to find the LCM is to keep adding the largest number to itself until the other number(s) divide into the sum evenly. To this end, we add 24 to itself: $24 + 24 = 48$. But 18 does not divide evenly into 48, so we add 24 to itself again: $24 + 24 + 24 = 72$. Now, 18 does divide evenly into 72 ($72/18 = 4$), so the LCM of 18 and 24 is 72.

$$\frac{18}{6} = 3 \qquad \frac{24}{6} = 4 \qquad \text{6 is the greatest common divisor}$$

$$\frac{LCM}{GCD} = \frac{72}{6} = 12$$

Answer: (D) 12

Solution:

2. The Syracuse Panthers baseball team won 54 games and lost 16 games. If they want to get an 80% winning average, how many more games do they have to win without any losses?

(A) 8   (B) 10   (C) 12   (D) 14   (E) 9

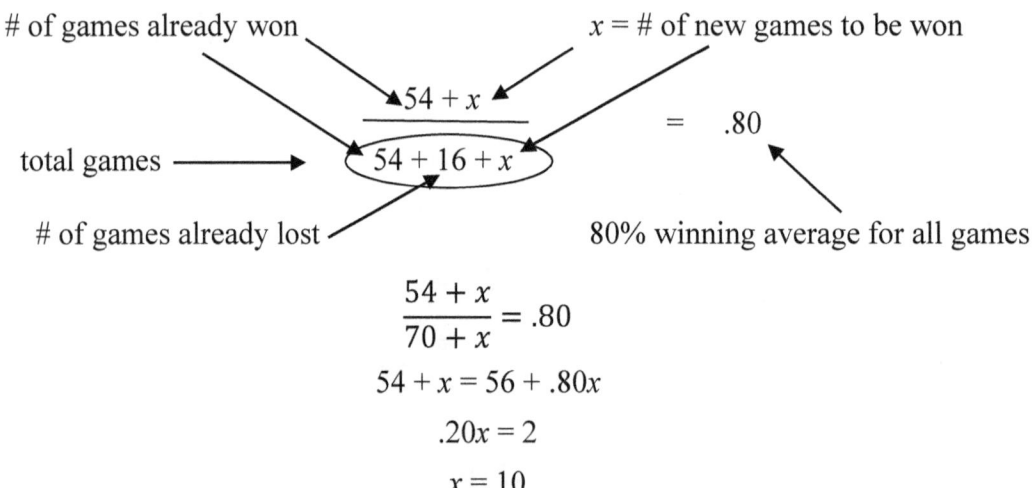

$$\frac{54 + x}{70 + x} = .80$$

$$54 + x = 56 + .80x$$

$$.20x = 2$$

$$x = 10$$

Answer: (B) 10

Test 1

3. Use the angle difference formula for sines to find the sin of 15°.

$$\sin(A - B) = \sin A \cos B - \cos A \sin B$$

(A) $\dfrac{\sqrt{2} - 3}{2}$ (B) $\dfrac{2\sqrt{3} + 1}{3\sqrt{2} - 2}$ (C) $\dfrac{3\sqrt{2} - 2}{2\sqrt{3} - 1}$ (D) $\dfrac{\sqrt{3} + 2}{2\sqrt{3}}$ (E) $\dfrac{\sqrt{3} - 1}{2\sqrt{2}}$

4. What is the domain of the function log $x = y$?

If we just write log x, the base is understood to be 10, so we could write the function as $\log_{10} x = y$. Then change to exponential form.

(A) $x < 0$ (B) $x > 0$ (C) $-\infty < x < 0$ (D) $-\infty < x < \infty$ (E) $x \geq 0$

3. Using the angle difference formula for sines, find the sin of 15°.

(A) $\dfrac{\sqrt{2}-3}{2}$ (B) $\dfrac{2\sqrt{3}+1}{3\sqrt{2}-2}$ (C) $\dfrac{3\sqrt{2}-2}{2\sqrt{3}-1}$ (D) $\dfrac{\sqrt{3}+2}{2\sqrt{3}}$ (E) $\dfrac{\sqrt{3}-1}{2\sqrt{2}}$

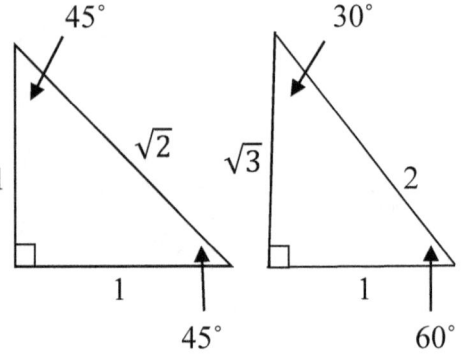

$\sin(A - B) = \sin A \cdot \cos B - \cos A \cdot \sin B$
$\sin(45° - 30°) = \sin 45° \cdot \cos 30° - \cos 45° \cdot \sin 30°$

$$\sin 15° = \dfrac{1}{\sqrt{2}} \cdot \dfrac{\sqrt{3}}{2} - \dfrac{1}{\sqrt{2}} \cdot \dfrac{1}{2} = \dfrac{\sqrt{3}}{2\sqrt{2}} - \dfrac{1}{2\sqrt{2}} = \dfrac{\sqrt{3}-1}{2\sqrt{2}}$$

Answer: (E) $\dfrac{\sqrt{3}-1}{2\sqrt{2}}$

4. What is the domain of the function $\log x = y$?

(A) $x < 0$ (B) $x > 0$ (C) $-\infty < x < 0$ (D) $-\infty < x < \infty$ (E) $x \geq 0$

The base of the common log is 10:

$$\log_{10} x = y$$

Recall that a log is the power to which you raise the base (10) to get the number you are taking the log of (x). Applying this definition yields

$$10^y = x$$

x	.000...1	.01	.1	10	100	1000...
y	$-\infty$	-2	-1	1	2	∞

So, x could be any real number greater than zero.

Answer: (B) $x > 0$

5. If the center of a circle is located at (3, 4) and its radius is 5, which of the following coordinates is located on the circumference of the circle?

Use the formula $(x - h)^2 + (y - k)^2 = r^2$, where (h, k) is the center of the circle, (x, y) are the coordinates of a point on the circumference and r is the radius.

(A) (5, 2)   (B) (6, 4)   (C) (6, 8)   (D) (5, 4)   (E) (6, 2)

6. In the equation $\sqrt[3]{a} + b^3 = 220$, if $b = 6$, what is the value of a?

(A) 28   (B) 36   (C) 54   (D) 42   (E) 64

To find the value of a, substitute 6 for b.

5. If the center of a circle is located at (3, 4) and its radius is 5, which of the following coordinates is located on the circumference of the circle?

(A) (5, 2)   (B) (6, 4)   (C) (6, 8)   (D) (5, 4)   (E) (6, 2)

Plug the choices into the equation for the circle to see which one satisfies the equation:

$(x - h)^2 + (y - k)^2 = r^2$

For Choice (C), we get

$(6 - 3)^2 + (8 - 4)^2 = 5^2$

$3^2 + 4^2 = 5^2$

$9 + 16 = 25$

Hence, Choice (C) satisfies the equation.

Answer: (C) (6, 8)

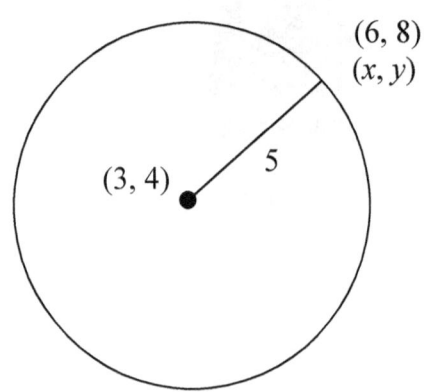

6. In the equation $\sqrt[3]{a} + b^3 = 220$, if $b = 6$, what is the value of a?

(A) 28   (B) 36   (C) 54   (D) 42   (E) 64

$$\sqrt[3]{a} + b^3 = 220$$

Plugging $b = 6$ into this equation yields

$$\sqrt[3]{a} + 6^3 = 220$$

$$\sqrt[3]{a} + 216 = 220$$

$$\sqrt[3]{a} = 4$$

$$a^{1/3} = 4$$

Raise both sides to the 3rd power. $\left(a^{1/3}\right)^3 = 4^3$

$$a = 64$$

Answer: (E) 64

7. A 747 airplane is fully refueled before takeoff. The 747 uses approximately 600 gallons of fuel per minute while flying at the speed of 600 mph and approximately 500 gallons per minute while flying at the speed of 500 mph. The plane flies for 3 hours at the rate of 600 mph and then slows down for the last 2 hours to 500 mph. On this entire flight what was the average number of gallons per mile used by the airplane?

(A) 25 (B) 50 (C) 60 (D) 45 (E) 55

Determine the *total* number of gallons used for the flight and the *total* mileage. Then divide the number of gallons used by the number of miles.

8. The table below lists the cost of manufacturing batteries. What is the <u>percent</u> decline in the cost per battery between manufacturing 1,500 batteries and manufacturing 6,500 batteries?

Number of batteries	Cost per battery
500–1,000	$.64
1,001–2,000	.60
2,001–5,00	.58
5,001–10,000	.57
10,001+	.56

(A) 10% (B) 9% (C) 8% (D) 5% (E) 6%

Use the cost per battery for manufacturing 1,500 batteries as a base from which you can determine the percent drop.

ACT Math Personal Tutor

7. A 747 airplane is fully refueled before takeoff. The 747 uses approximately 600 gallons of fuel per minute while flying at the speed of 600 mph and approximately 500 gallons per minute while flying at the speed of 500 mph. The plane flies for 3 hours at the rate of 600 mph and then slows down for the last 2 hours to 500 mph. On this entire flight what was the average number of gallons per mile used by the airplane?

(A) 25   (B) 50   (C) 60   (D) 45   (E) 55

At 600 mph: 600 gallons per minute × 60 minutes × 3 hours = 108,000 gallons
At 500 mph, 500 gallons per minute × 60 minutes × 2 hours = 60,000 gallons
                                                            168,000 gallons

total mileage: 600 mph × 3 hours = 1,800 miles
              + 500 mph × 2 hours = 1,000 miles
                                     2,800 miles

$$\frac{168{,}000 \text{ gallons}}{2{,}800 \text{ miles}} = 60 \text{ gallons per mile}$$

Answer: (C) 60

8. The table below lists the cost of manufacturing batteries. What is the <u>percent</u> decline in the cost per battery between manufacturing 1,500 batteries and manufacturing 6,500 batteries?

Number of batteries	Cost per battery
500–1,000	$.64
1,001–2,000	.60
2,001–5,00	.58
5,001–10,000	.57
10,001+	.56

(A) 10%   (B) 9%   (C) 8%   (D) 5%   (E) 6%

The cost for manufacturing each battery dropped from $0.60 to $0.57 — a $.03 drop. To find the percent drop, divide the change in cost by the original cost:

$$\frac{Change\ in\ cost}{Original\ cost} = \frac{\$.03}{\$.60} = .05 = 5\%$$

Answer: (D) 5%

9. If $a = 121$ and $b = a + 4$, find the value of $b^{2/3}$.

(A) 25 (B) 10 (C) 12 (D) 21 (E) 14

Substitute $a + 4$ for b in the expression $b^{2/3}$. Find the cube root and then square it.

10. Solve for x in the equation $\sin^2 x + 2\sin x + 1 = 0$, where $0° \leq x \leq 360°$.

(A) 300° (B) 60° (C) 120° (D) 30° (E) 270°

Let $y = \sin x$ in the equation.

9. If $a = 121$ and $b = a + 4$, find the value of $b^{2/3}$.

(A) 25   (B) 10   (C) 12   (D) 21   (E) 14

$a = 121$:
$$\begin{aligned} b^{2/3} &= (a+4)^{2/3} \\ &= (121+4)^{2/3} \\ &= (125)^{2/3} \\ &= 5^2 \\ &= 25 \end{aligned}$$

Answer: (A) 25

10. Solve for x in the equation $\sin^2 x + 2\sin x + 1 = 0$, where $0° \leq x \leq 360°$.

(A) 300°   (B) 60°   C) 120°   (D) 30°   (E) 270°

Let $y = \sin x$:
$$\begin{aligned} \sin^2 x + 2\sin x + 1 &= 0 \\ y^2 + 2y + 1 &= 0 \\ (y+1)(y+1) &= 0 \\ y = -1 \text{ or } y &= -1 \end{aligned}$$

Now, replacing y with $\sin x$, yields $\sin x = -1$.

From the graph, $\sin x = -1$ when $x = 270°$.

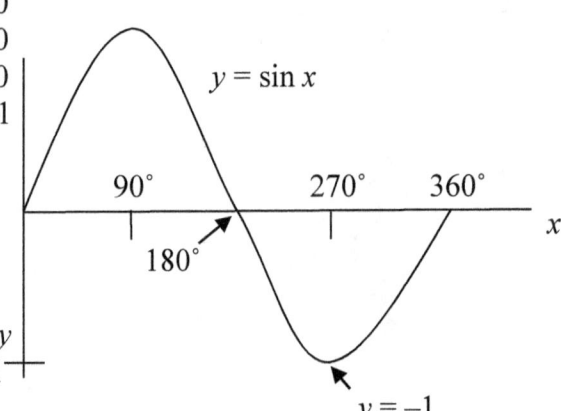

Answer: (E) 270

11. How many degrees are there in an exterior angle of a regular octagon?

(A) 110° (B) 45° (C) 140° (D) 135° (E) 60°

Total number of degrees in a polygon is $(n-2) \cdot 180°$, where $n =$ the number of sides of the polygon.

12. If each line represents one unit, select the equation represented by this parabola.

(A) $y = x^2 + 2x - 4$ (B) $y = 2x^2 + 3x + 2$ (C) $x = y^2 - 2y - 9$ (D) $y = x^2 - x - 6$
(E) $y = x^2 - 2x + 6$

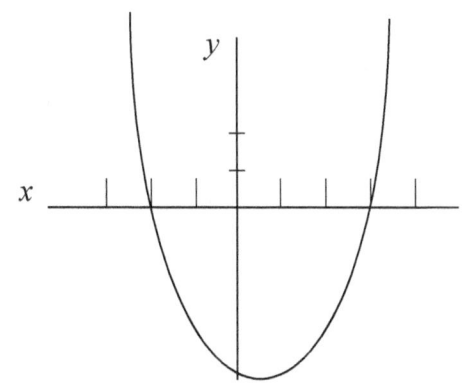

The parabola crosses the x-axis at the points $x = -2$ and $x = 3$.

11. How many degrees are there in an external angle of a regular octagon?

(A) 110°   (B) 45°   (C) 140°   (D) 135°   (E) 60°

Since $(n-2) \cdot 180°$ gives the total number of degrees in a polygon, $\dfrac{(n-2)180°}{n}$ gives the number of degrees in each *interior* angle of the regular polygon

interior ∠ = 135°   exterior ∠ = 45°

$$\dfrac{(8-2)180°}{8} = 135°$$

Answer: (B) 45°

12. If each line represents one unit, select the equation represented by this parabola.

(A) $y = x^2 + 2x - 4$ (B) $y = 2x^2 + 3x + 2$ (C) $x = y^2 - 2y - 9$ (D) $y = x^2 - x - 6$
(E) $y = x^2 - 2x + 6$

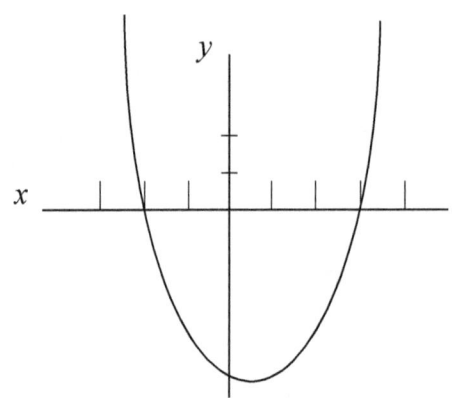

The parabola crosses the x-axis at the points $x = -2$ and $x = 3$, so the equation is

$$(x + 2)(x - 3) = 0$$
$$x^2 - 3x + 2x - 6 = 0$$
$$x^2 - x - 6 = 0$$

Answer: (D) $y = x^2 - x - 6$

13. If $B = \left|\dfrac{x-2}{x^2-4}\right|$, $-1 \leq x \leq 5$, $x \neq \pm 2$, choose one of the following integers to find an x that results in the smallest value for B.

(A) –1　　(B) 0　　(C) 3　　(D) 4　　(E) 5

Factor and cancel. Then see which value of x results in the smallest B.

14. If $b = 6a$ and $c = 2b$, simplify the ratio of $\dfrac{b^2}{3abc}$ in terms of a.

(A) 1/6a　　(B) $3a^2$　　(C) 3/5a　　(D) 3/4a　　(E) 2a/5

Substitute $6a$ for b and $2b$ for c. Then reduce.

13. If $B = \left|\dfrac{x-2}{x^2-4}\right|$, $-1 \leq x \leq 5$, $x \neq \pm 2$, choose one of the following integers to find an x that results in the smallest value for B.

(A) –1   (B) 0   (C) 3   (D) 4   (E) 5

$$B = \left|\dfrac{\cancel{x-2}}{(x+2)\cancel{(x-2)}}\right| = \left|\dfrac{1}{x+2}\right|$$

The largest value for x in the denominator will result in the smallest value for B, so the answer is 5.

Answer: (E) 5

14. If $b = 6a$ and $c = 2b$, simplify the ratio of $\dfrac{b^2}{3abc}$ in terms of a.

(A) 1/6a   (B) $3a^2$ (C) 3/5a   (D) 3/4a   (E) 2a/5

$$\dfrac{b^2}{3abc} =$$

$b = 6a, c = 2b$:
$$\dfrac{(6a)^2}{3a(6a)(2b)} =$$

$b = 6a$:
$$\dfrac{(6a)^2}{3a(6a)(2[6a])} =$$

$$\dfrac{36a^2}{216a^3} =$$

$$\dfrac{1}{6a}$$

Answer: (A) 1/6a

15. In a study of people's color preferences, 40 people chose black. The percentages of the rest of the responses are indicated in the circle diagram shown. How many preferred green?

(A) 48
(B) 40
(C) 64
(D) 60
(E) 52

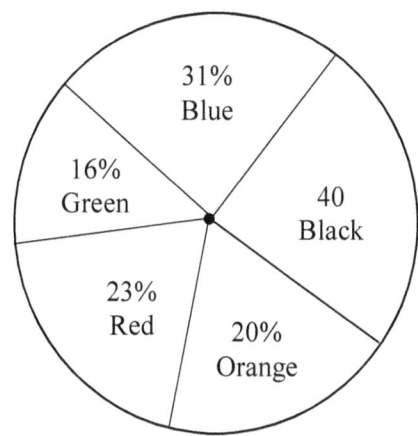

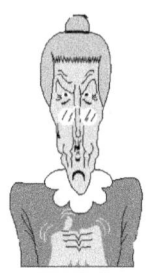

Set up a proportion between the number of black choices to the % black compared to the number of green choices to the % green.

15. In a study of people's color preferences, 40 people chose black. The percentages of the rest of the responses are indicated in the circle diagram shown. How many preferred green?

(A) 48
(B) 40
(C) 64
(D) 60
(E) 52

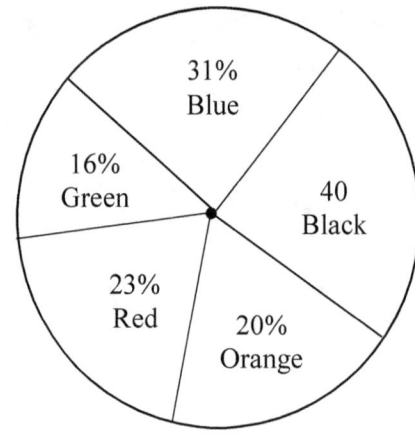

First find out what percent preferred black:

$$100\% - \text{Blue} - \text{Green} - \text{Red} - \text{Orange} =$$

$$100\% - 31\% - 16\% - 23\% - 20\% = 10\%$$

Let x = the number of green preferences.

Now, we'll do a proportion:

$$\frac{\text{\# black}}{\text{\% black}} = \frac{\text{\# green}}{\text{\% green}}$$

$$\frac{40}{.10} = \frac{x}{.16}$$

$$.10x = 6.4$$

$$x = 64$$

Answer: (C) 64

16. A full box of ball bearings weighs 21 pounds. The box plus 1/3 of the ball bearings weighs 9 pounds. How much does the box alone weigh?

(A) 4 pounds   (B) 3 pounds   (C) 5 pounds   (D) 6 pounds   (E) 2 pounds

Set up two simultaneous equations.
Let x = weight of the box alone.
Let y = the total weight of the ball bearings alone.
Let $y/3$ = the weight of 1/3 of the ball bearings.

16. A full box of ball bearings weighs 21 pounds. The box plus 1/3 of the ball bearings weighs 9 pounds. How much does the box alone weigh?

(A) 4 pounds   (B) 3 pounds   (C) 5 pounds   (D) 6 pounds   (E) 2 pounds

Let x = the weight of the empty box.
Let y = the weight of the ball bearings alone.
Let $y/3$ = the weight of 1/3 of the ball bearings.

(1) One full box = (empty box) + (ball bearings) = 21 pounds.
(2) The box plus 1/3 of the ball bearings = 9 pounds.

$$(1)\ x + y = 21$$
$$(2)\ x + y/3 = 9$$

$$(1)\ x = 21 - y$$

$x = 21 - y$:
$$(2)\ (21 - y) + y/3 = 9$$

$$(2)\ -2y/3 = -12$$

$$y = 18$$

$$(1)\ x + y = 21$$
$$(1)\ x + 18 = 21$$
$$(1)\ x = 3$$

Answer: (B) 3 pounds

17. The diagram shown depicts 3 activities out of a 24-hour day. What percent of a 24-hour day did women older than 25 have for activities other than watching TV, studying or sleeping?

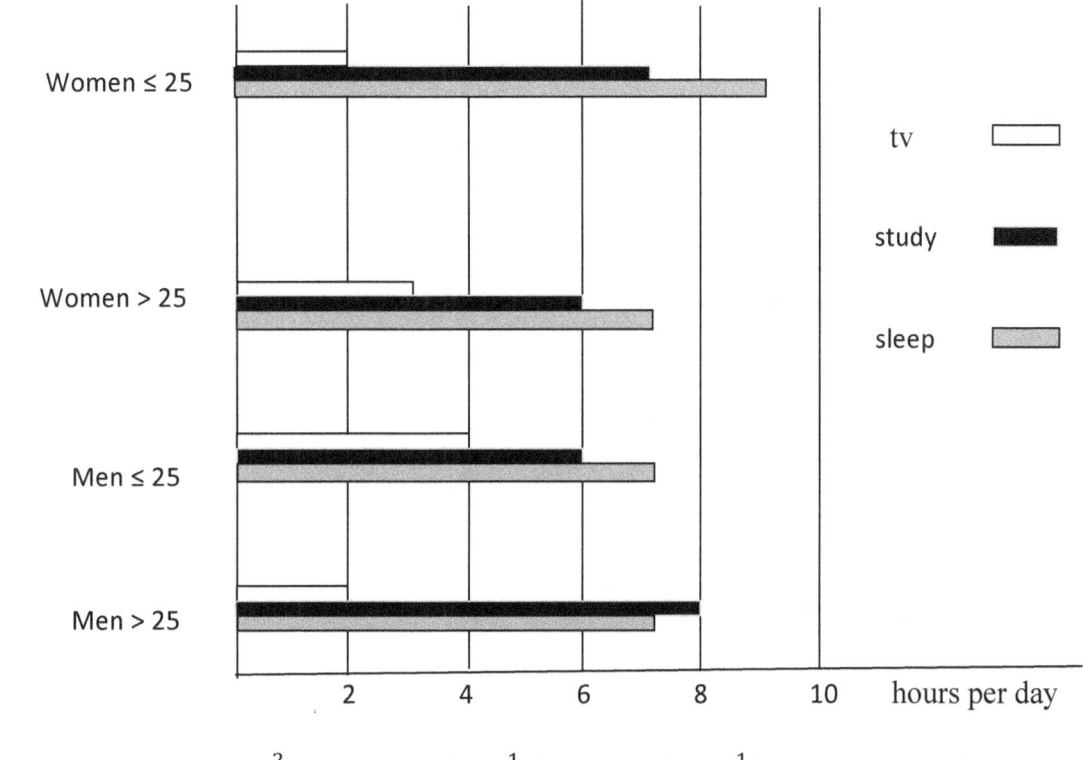

(A) 20% (B) $35\frac{2}{3}\%$ (C) $27\frac{1}{2}\%$ (D) $33\frac{1}{3}\%$ (E) 26%

Add up the hours women over 25 spent on the 3 activities. Using this total, subtract from 24 hours and then determine what % of a 24-hour day is spent on other activities.

17. The diagram shown depicts 3 activities out of a 24-hour day. What percent of a 24-hour day did women older than 25 have for activities other than watching TV, studying or sleeping?

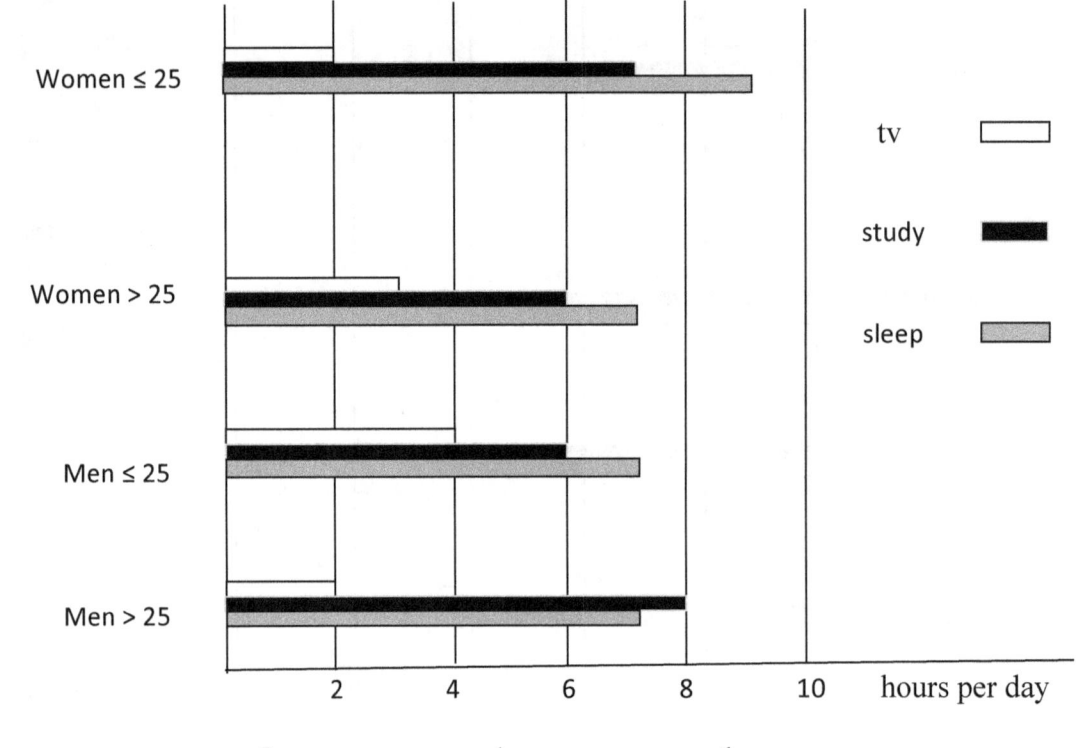

(A) 20% (B) $35\frac{2}{3}$% (C) $27\frac{1}{2}$% (D) $33\frac{1}{3}$% (E) 26%

The women over 25 spent 3 hours watching TV, 6 hours studying and 7 hours sleeping. That left them 24 – 3 – 6 – 7 = 8 hours for other activities.

$$\frac{8}{24} = \frac{1}{3} = 33\frac{1}{3}\%$$

Answer: (D) $33\frac{1}{3}$%

18. ABCD is an isosceles trapezoid, with AB = CD. AG ⊥ EF. AD = 4, BC = 10, EF = 8 and AG = 6. E and F are the midpoints of AB and CD, respectively. The height of ABCD is 12. Find the area of trapezoid EFCB.

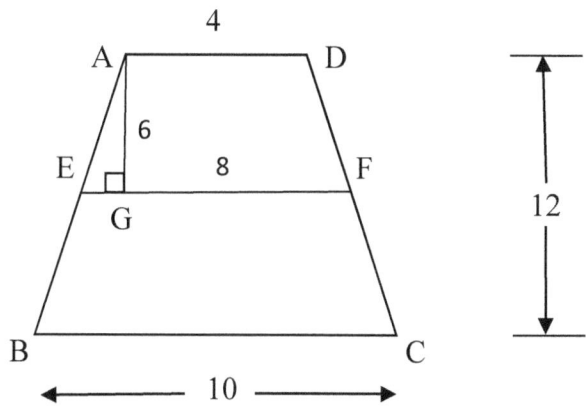

(A) 42   (B) 48   (C) 36   (D) 78   (E) 60

The formula for the area of a trapezoid is

$$A = \frac{1}{2}h(b_1 + b_2)$$

where A = area, h = height, b_1 = one base and b_2 = the other base.

19. At what point does the inverse function of $y = 5x + 3$ intersect the x-axis?

(A) (5, 0)   (B) (4, 0)   (C) (2, 0)   (D) (3, 0)   (E) (0, 3)

The inverse of $y = 5x + 3$ is $x = 5y + 3$.

18. ABCD is an isosceles trapezoid, with AB = CD. AG ⊥ EF. AD = 4, BC = 10, EF = 8 and AG = 6. E and F are the midpoints of AB and CD, respectively. The height of ABCD is 12. Find the area of trapezoid EFCB.

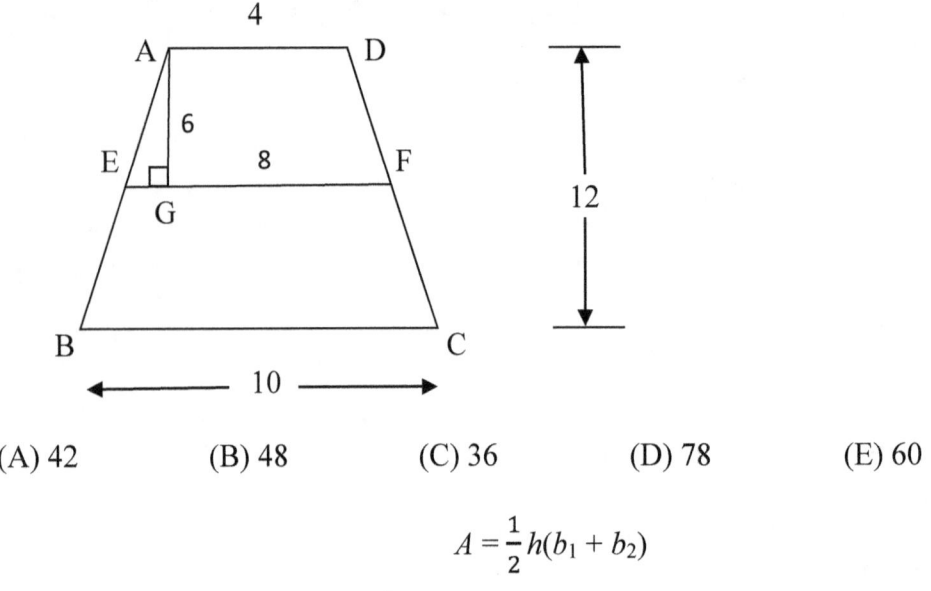

(A) 42   (B) 48   (C) 36   (D) 78   (E) 60

$$A = \frac{1}{2}h(b_1 + b_2)$$

For trapezoid ABCD, Area = $\frac{1}{2}(h)(b_1 + b_2)$ = (1/2)(12)(4 + 10) =   84

Subtract Area of trapezoid ADFE = $(\frac{1}{2})(h)(b_1 + b_2)$ = (1/2)(6)(4 + 8) = 36
$$\overline{}$$
                                                                                    48

Answer: (B) 48

19. At what point does the inverse function of $y = 5x + 3$ intersect the x-axis?

(A) (5, 0)   (B) (4, 0)   (C) (2, 0)   (D) (3, 0)   (E) (0, 3)

$y = 5x + 3$

Forming the inverse yields $x = 5y + 3$.

Solving for y yields $y = \frac{x-3}{5}$.

The graph of the inverse function intersects the x-axis at $y = 0$.

So, $y = \frac{x-3}{5} = 0$.

Hence, $x = 3$

The graph of the f^{-1} intersects the x-axis at (3, 0).

Answer: (D) (3, 0)

20. In triangle ABC shown, angle ABC is 90°, BD ⊥ AC and the measure of angle C is 60°. CD = 4. Find the height of triangle ABC.

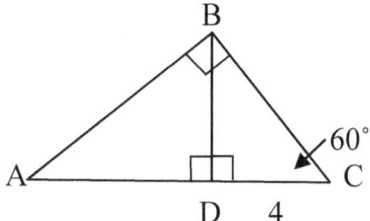

(A) $3\sqrt{3}$ (B) 6   (C) $5\sqrt{3}$
(D) $4\sqrt{3}$ (E) $4\sqrt{2}$

Draw a 30°–60°–90° triangle adjacent to the given triangle, so that we can set up a proportion in order to find BD.

21. Ziggy wants to arrange a penny, a nickel, a dime, a quarter and a half-dollar in a row in different ways. How many different arrangements can he make?

(A) 80   (B) 90   (C) 100   (D) 110   (E) 120

There are five positions to be filled. Once a coin is chosen, there is one less coin to choose from to fill the next position.

20. In triangle ABC shown, angle ABC is 90°, BD ⊥ AC and the measure of angle C is 60°. CD = 4. Find the height of triangle ABC.

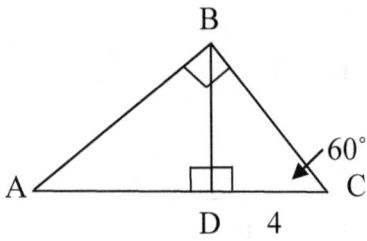

(A) $3\sqrt{3}$ (B) 6 (C) $5\sqrt{3}$
(D) $4\sqrt{3}$ (E) $4\sqrt{2}$

Triangles BDC and GFE are similar. We can find the height using proportions.

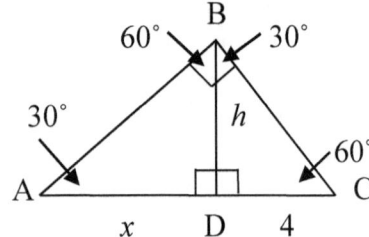

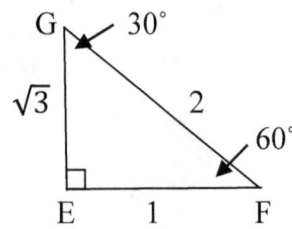

Forming the proportion yields

$$\frac{4}{1} = \frac{h}{\sqrt{3}}$$

$$h = 4\sqrt{3}$$

Answer: (D) $4\sqrt{3}$

21. Ziggy wants to arrange a penny, a nickel, a dime, a quarter and a half-dollar in a row in different ways. How many different arrangements can he make?

(A) 80 (B) 90 (C) 100 (D) 110 (E) 120

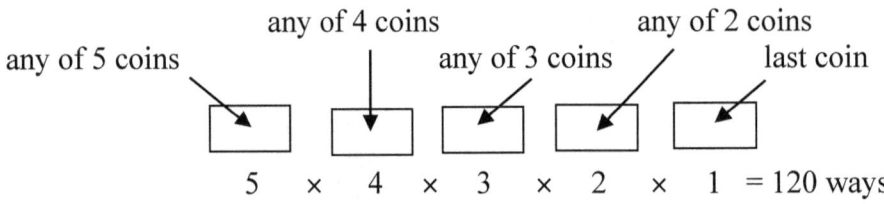

Answer: (E) 120

22. Which algebraic expression represents the 9th term of an arithmetic sequence whose first two terms are *a* and *b*?

(A) $a + 8(b - a)$ (B) $a + 9b$ (C) $9(a - b)$ (D) $8(a - b)$ (E) $a + 9(b - a)$

The *n*th term of an arithmetic sequence is found by using the formula $a_n = a_1 + (n - 1)d$, where a_n is the *n*th term, a_1 is the first term, *n* = the number of terms and *d* = the common difference of successive terms.

23. If $\dfrac{c}{d}$ is 8% of *e* and *d* = 4, find the ratio of *e* : *c*.

(A) 1 : (.32)   (B) 3 : (.20)   (C) 1 : (.12)   (D) 2 : (.45)   (E) 2 : (.25)

Translate the clause "*c*/*d* is 8% of *e*" into an equation, solve for *c*, and then find the ratio *e* : *c*.

22. Which algebraic expression represents the 9th term of an arithmetic sequence whose first two terms are a and b?

(A) $a + 8(b - a)$ (B) $a + 9b$ (C) $9(a - b)$ (D) $8(a - b)$ (E) $a + 9(b - a)$

$$a_n = a_1 + (n - 1)d$$

Since a and b are the first two terms of the sequence, they are successive terms and the common difference is $d = b - a$. Substituting this into the nth term formula gives

$$a_n = a_1 + (n - 1)(b - a)$$

Since we are looking for the 9th term, replace n with 9:

$$a_9 = a + (9 - 1)(b - a)$$

$$a_9 = a + 8(b - a)$$

Answer: (A) $a + 8(b - a)$

23. If $\dfrac{c}{d}$ is 8% of e and $d = 4$, find the ratio of $e : c$.

(A) $1 : (.32)$ (B) $3 : (.20)$ (C) $1 : (.12)$ (D) $2 : (.45)$ (E) $2 : (.25)$

$$\frac{c}{d} = 8\%e$$
$$\frac{c}{d} = .08e$$
$$c = .08de$$

$d = 4$:
$$c = .08(4)e$$
$$c = .32e$$
$$\frac{e}{c} = \frac{e}{.32e} = \frac{1}{.32}$$

Writing this result in proportion notation gives

$$e : c = 1 : (.32)$$

Answer: (A) $1 : (.32)$

24. If
$$f(x) = \frac{4|x|}{\frac{2}{3}x^{-2}}$$
find the value of $f(-3)$.

(A) 162 (B) 58 (C) 146 (D) 172 (E) 86

Substitute −3 for x.

25. Lines AB and CD intersect at E. Angle AED measures 96°. Angle DEF measures 14° more than angle FEB. How many degrees are there in angle DEF?

(A) 35° (B) 56° (C) 44° (D) 38° (E) 49°

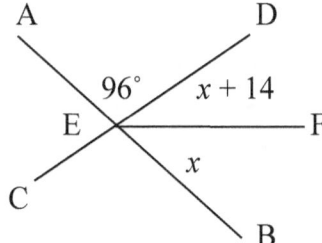

The measure of ∠AEB = 180°.

24. If
$$f(x) = \frac{4|x|}{\frac{2}{3}x^{-2}}$$
find the value of $f(-3)$.

(A) 162 (B) 58 (C) 146 (D) 172 (E) 86

$$f(-3) = \frac{4|-3|}{\frac{2}{3}(-3)^{-2}} = \frac{4(3)}{\frac{2}{3}\cdot\frac{1}{9}} = \frac{12}{\frac{2}{27}} = 12\cdot\frac{27}{2} = 162$$

Answer: (A) 162

25. Lines AB and CD intersect at E. Angle AED measures 96°. Angle DEF measures 14° more than angle FEB. How many degrees are there in angle DEF?

(A) 35° (B) 56° (C) 44° (D) 38° (E) 49°

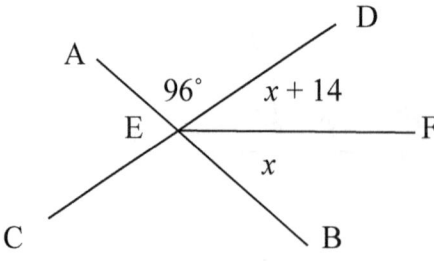

$m\angle AEB = 180°$
$96° + x + (x + 14) = 180°$
$2x + 110° = 180°$
$2x = 70°$
$x = 35°$
$x + 14° = 49°$

$\angle DEF = 49°$.

Answer: (E) 49°

26. If $1 \leq |x| \leq 4$ and $3 \leq |y| \leq 6$, which of the following inequalities is FALSE?

(A) $4 \leq |x| + |y| \leq 10$ (B) $8 \leq 2|x| + 2|y| \leq 20$ (C) $3 \leq |x| \cdot |y| \leq 24$

(D) $-1 \leq |y| - |x| \leq 5$ (E) $-5 \leq |x| - |y| \leq 0$

Add, subtract, and multiply the two given inequalities.

27. If $f^{-1}(x) = \dfrac{x+4}{2}$, find $f(3)$.

(A) 3 (B) 5 (C) 2 (D) 6 (E) 4

Let $y = (x + 4)/2$. Switch x and y to find $f(x)$ and then substitute 3 for x.

ACT Math Personal Tutor

26. If $1 \leq |x| \leq 4$ and $3 \leq |y| \leq 6$, which of the following inequalities is FALSE?

(A) $4 \leq |x| + |y| \leq 10$ (B) $8 \leq 2|x| + 2|y| \leq 20$ (C) $3 \leq |x| \cdot |y| \leq 24$

(D) $-1 \leq |y| - |x| \leq 5$ (E) $-5 \leq |x| - |y| \leq 0$

Choice (A): Add the given inequalities as is

$$1 \leq |x| \leq 4$$
$$(+) \quad 3 \leq |y| \leq 6$$
$$1 + 3 \leq |x| + |y| \leq 4 + 6$$
$$4 \leq |x| + |y| \leq 10$$

Choice (B): Multiply each inequality by 2 and then add:

$$2 \leq 2|x| \leq 8$$
$$(+) \quad 6 \leq 2|y| \leq 12$$
$$2 + 6 \leq 2|x| + 2|y| \leq 8 + 12$$
$$8 \leq 2|x| + 2|y| \leq 20$$

Choice (C): Multiply the inequalities as is:

$$1 \leq |x| \leq 4$$
$$(\cdot) \quad 3 \leq |y| \leq 6$$
$$1 \cdot 3 \leq |x| \cdot |y| \leq 4 \cdot 6$$
$$3 \leq |x| \cdot |y| \leq 24$$

Choice (D): Multiply the inequality $1 \leq |x| \leq 4$ by -1: $-1 \geq -|x| \geq -4$. Rewriting this inequality in the natural order (smallest number first) gives $-4 \leq -|x| \leq -1$. Now, add this inequality to given inequality $3 \leq |y| \leq 6$:

$$3 \leq |y| \leq 6$$
$$(+) \quad -4 \leq -|x| \leq -1$$
$$3 + (-4) \leq |y| + (-|x|) \leq 6 + (-1)$$
$$-1 \leq |y| - |x| \leq 5$$

This proves that choices (A), (B), (C), and (D) are all true. Hence, by process of elimination, the answer is (E) $-5 \leq |x| - |y| \leq 0$.

27. If $f^{-1}(x) = \dfrac{x+4}{2}$, find $f(3)$.

(A) 3   (B) 5   (C) 2   (D) 6   (E) 4

Replace $f^{-1}(x)$ with y:

$$y = \frac{x+4}{2}$$

Form the inverse by interchanging the x and y variables:

$$x = \frac{y+4}{2}$$

Solving for y yields

$$y = 2x - 4$$

Or with function notation:

$$f(x) = 2x - 4$$
$$f(3) = 2(3) - 4$$
$$f(3) = 2$$

Answer: (C) 2

28. If $x^{n-2} = x^3 \cdot x^{2n-8}$, find the value of n.

(A) 3 (B) 4 (C) 2 (D) 5 (E) 6

> For any real numbers a, b and c, $a^b \cdot a^c = a^{b+c}$

29. If there are 2 red balls, 4 blue balls, 6 white balls and a certain number of black balls in a container and the odds of selecting a black ball are 1 out of 5, how many black balls are there?

(A) 5 (B) 3 (C) 4 (D) 2 (E) 6

> Let b = the number of black balls. Add b to the total of all the other balls. Then set the ratio of black balls to the total number of balls equal to the odds of selecting a black ball.

28. If $x^{n-2} = x^3 \cdot x^{2n-8}$, find the value of n.

(A) 3　　　　(B) 4　　　　(C) 2　　　　(D) 5　　　　(E) 6

$$x^{n-2} = x^3 \cdot x^{2n-8}$$

$$x^{n-2} = x^{2n-8+3}$$

$$x^{n-2} = x^{2n-5}$$

Since the bases (x) are equal, the exponents must be equal:

$$n - 2 = 2n - 5$$

$$n = 3$$

Answer: (A) 3

29. If there are 2 red balls, 4 blue balls, 6 white balls and a certain number of black balls in a container and the odds of selecting a black ball are 1 out of 5, how many black balls are there?

(A) 5　　　　(B) 3　　　　(C) 4　　　　(D) 2　　　　(E) 6

Let b = number of black balls.
Total number of balls = $2 + 4 + 6 + b = 12 + b$.

Odds of selecting a black ball = $\dfrac{\text{number of black balls}}{\text{total number of balls}}$

$$\frac{1}{5} = \frac{b}{12 + b}$$

$$5b = 12 + b$$

$$4b = 12$$

$$b = 3$$

Answer: (B) 3

30. Given rhombus ABCD, diagonals AC and BD intersect at E. If the measure of angle BCF is 76°, find the measure of angle DBC.

(A) 38°   (B) 26°   (C) 56°   (D) 42°   (E) 29°

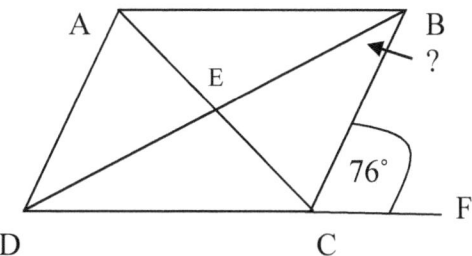

The diagonals of a rhombus are perpendicular to each other and the sides are all congruent.

30. Given rhombus ABCD, diagonals AC and BD intersect at E. If the measure of angle BCF is 76°, find the measure of angle DBC.

(A) 38° (B) 26° (C) 56° (D) 42° (E) 29°

The diagonals of a rhombus are perpendicular to each other.

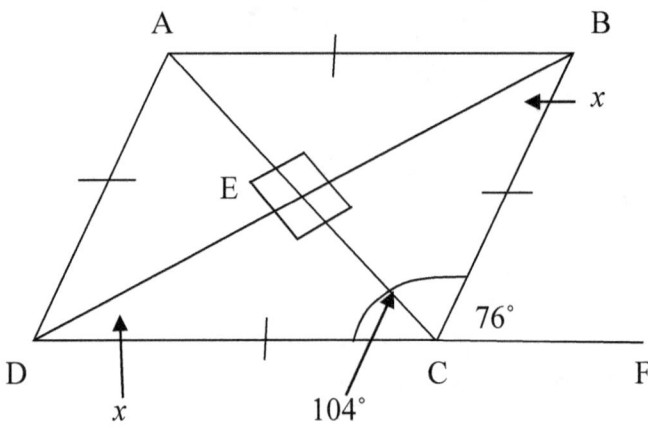

Angle DCB is supplementary to angle BCF, so $m\angle DCB = 180° - 76° = 104°$.

A rhombus is a parallelogram with four congruent sides, so triangle BCD is isosceles with two equal base angles (the x's in the figure).

Sum of the measures of the angles in a triangle equals 180°.
$$x + x + 104 = 180°$$
$$2x = 76°$$
$$x = 38°$$

Answer: (A) 38°

31. If a square is inscribed inside a circle, what is the ratio of the area of the circle to the area of the square?

(A) $3\pi/2$ (B) $2/\pi$ (C) $3/\pi$ (D) $\pi/3$ (E) $\pi/2$

A diagonal of the square is equal to the diameter of the circle. Use the Pythagorean Theorem to find a side of the square and then you can compare areas.

31. If a square is inscribed inside a circle, what is the ratio of the area of the circle to the area of the square?

(A) $3\pi/2$ (B) $2/\pi$ (C) $3/\pi$ (D) $\pi/3$ (E) $\pi/2$

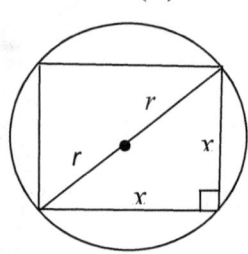

The sides of a square are equal, so we can let each side equal x.

The hypotenuse of the triangle is equal to twice the radius, so the hypotenuse is $2r$.

By the Pythagorean Theorem,

$$x^2 + x^2 = (2r)^2$$
$$2x^2 = 4r^2$$
$$x^2 = 2r^2$$
$$x = r\sqrt{2}$$

Hence, the area of the square is

$$r\sqrt{2} \cdot r\sqrt{2} = 2r^2$$

Now, the area of circle is πr^2.

So, the ratio of the area of the circle to the area of the square is

$$\frac{\text{Area of circle}}{\text{Area of square}} = \frac{\pi r^2}{2r^2} = \frac{\pi}{2}$$

Answer: (E) $\pi/2$

32. In a survey of twenty-six men and twenty-nine women, ten out of eighteen of those with blonde hair were men. Nine out of nineteen with red hair were women. All the rest were brunettes. How many women were brunettes?

(A) 12   (B) 8   (C) 9   (D) 10   (E) 6

Set up a table with rows of men and women and columns of hair color.

33. When full, a right circular cylinder can contains 55π cubic feet of water and has a radius $\sqrt{5}$. However, it is filled to only a height of 6 feet. How much height is left above the water?

(A) 6 ft.   (B) 4 ft.   (C) 7 ft.   (D) 3 ft.   (E) 5 ft.

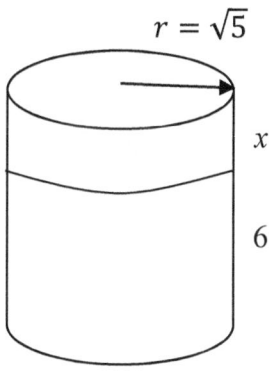

Volume of a cylinder = $\pi r^2 h$. Let x = the empty space above the water, and let the total height of the cylinder = $x + 6$.

32. In a survey of twenty-six men and twenty-nine women, ten out of eighteen of those with blonde hair were men. Nine out of nineteen with red hair were women. All the rest were brunettes. How many women were brunettes?

(A) 12   (B) 8   (C) 9   (D) 10   (E) 6

	Blondes	+	Redheads	+	Brunettes	=	Total
Men	10	+	10	+	6		26
Women	8	+	9	+	12	=	29
Total	18	+	19	+	18	=	55

Answer: (A) 12

33. When full, a right circular cylinder can contains 55π cubic feet of water and has a radius $\sqrt{5}$. However, it is filled to only a height of 6 feet. How much height is left above the water?

(A) 6 ft.   (B) 4 ft.   (C) 7 ft.   (D) 3 ft.   (E) 5 ft.

radius, $r = \sqrt{5}$, height, $h = 6 + x$

$$\text{volume} = \pi r^2 h$$
$$55\pi = \pi(\sqrt{5})^2 (6 + x)$$
$$55\pi = 5\pi(6 + x)$$
$$55\pi = 30\pi + 5\pi x$$
$$25\pi = 5\pi x$$
$$x = 5$$

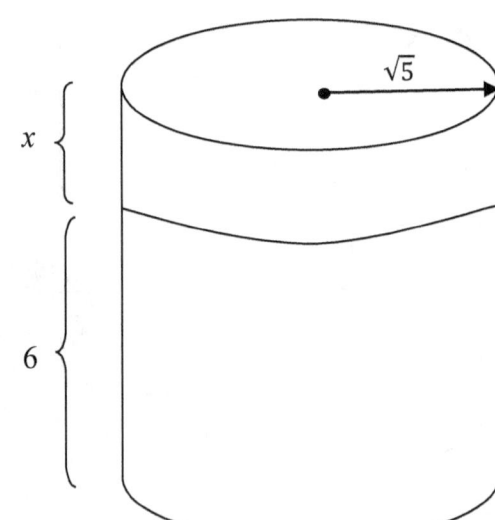

Answer: (E) 5 ft.

34.

$$A = \left\{\frac{n}{2} \,\middle|\, 2 < \sqrt{n} < 3 \text{ and } n \in Z, \text{ where Z is the set of positive integers}\right\}$$

Find set A.

(A) {2, 3.5, 4, 4.5} (B) {2.5, 3, 3.5, 4} (C) {3.5, 4.5, 5.5, 6}
(D) {4, 4.5, 5, 5.5} (E) {3, 4, 4.5, 5, 5.5}

Square all parts of the inequality $2 < \sqrt{n} < 3$ in order to find the range of values for n. Then substitute all values of n into the fraction $\frac{n}{2}$.

35. Simplify $\tan^2\theta \cdot \cos^2\theta$.

(A) $\cos^2\theta$ (B) 1   (C) $\sin\theta$ (D) $\cos\theta$ (E) $\sin^2\theta$

$$\tan^2\theta = \frac{\sin^2\theta}{\cos^2\theta}$$

34.

$$A = \left\{\frac{n}{2} \,\middle|\, 2 < \sqrt{n} < 3 \text{ and } n \in Z, \text{ where Z is the set of positive integers}\right\}$$

Find set A.

(A) {2, 3.5, 4, 4.5}   (B) {2.5, 3, 3.5, 4}   (C) {3.5, 4.5, 5.5, 6}
(D) {4, 4.5, 5, 5.5}   (E) {3, 4, 4.5, 5, 5.5}

$$2 < \sqrt{n} < 3$$
$$2 < n^{1/2} < 3$$

Now, square each term of the inequality, which will preserve the direction of the inequality because all the terms are positive:

$$2^2 < \left(n^{1/2}\right)^2 < 3^2$$
$$4 < n < 9$$

So, n is between 4 and 9, and $n \in Z$, where Z is the set of positive integers. That is, n is one of the integers 5, 6, 7, 8.

So, we'll select all the integers greater than 4 and less than 9 and divide each of them by 2 because we are looking for numbers of the form $n/2$:

$$\frac{5}{2} = 2.5, \quad \frac{6}{2} = 3, \quad \frac{7}{2} = 3.5, \quad \frac{8}{2} = 4$$

Answer: (B) {2.5, 3, 3.5, 4}

35. Simplify $\tan^2\theta \cdot \cos^2\theta$.

(A) $\cos^2\theta$ (B) 1   (C) $\sin\theta$ (D) $\cos\theta$ (E) $\sin^2\theta$

$$\tan^2\theta \cdot \cos^2\theta$$

$$\frac{(\sin\theta)^2}{(\cos\theta)^2} \cdot (\cos\theta)^2$$

Canceling $\cos^2\theta$ yields

$$\sin^2\theta$$

Answer: (E) $\sin^2\theta$

36. If the diagonal of a square is $d^{1/4}$, find a side of the square.

(A) $\dfrac{d^{1/8}}{2^{1/2}}$ (B) $\dfrac{d^{1/4}}{2^{1/2}}$ (C) $\dfrac{d^{2/3}}{2^{1/4}}$ (D) $\dfrac{d^{1/16}}{2^{1/2}}$ (E) $\dfrac{d^{1/2}}{2^{1/4}}$

Use the Pythagorean Theorem to find a side of the square.

37. If a represents an integer, what is the additive inverse of $a^2 - 2a$?

(A) $2a - a^2$ (B) $a^2 + 2a$ (C) $-2a + a^2$ (D) $a^2 - 2a$ (E) $-2a - a^2$

The additive inverse of a number, n, is another number, which when added to n results in zero.

36. If the diagonal of a square is $d^{1/4}$, find a side of the square.

(A) $\dfrac{d^{1/8}}{2^{1/2}}$ (B) $\dfrac{d^{1/4}}{2^{1/2}}$ (C) $\dfrac{d^{2/3}}{2^{1/4}}$ (D) $\dfrac{d^{1/16}}{2^{1/2}}$ (E) $\dfrac{d^{1/2}}{2^{1/4}}$

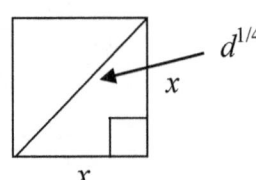

Use the Pythagorean Theorem: $a^2 + b^2 = c^2$, where a and b are the sides of a right triangle and c is the hypotenuse.

$$x^2 + x^2 = (d^{1/4})^2$$

$$2x^2 = d^{1/2}$$

$$x^2 = \frac{d^{1/2}}{2}$$

$$x = \left(\frac{d^{1/2}}{2}\right)^{1/2} = \frac{d^{1/4}}{2^{1/2}}$$

Answer: (B) $\dfrac{d^{1/4}}{2^{1/2}}$

37. If a represents an integer, what is the additive inverse of $a^2 - 2a$?

(A) $2a - a^2$ (B) $a^2 + 2a$ (C) $-2a + a^2$ (D) $a^2 - 2a$ (E) $-2a - a^2$

The additive inverse of a number, n, is another number, which when added to n results in zero.

$$(a^2 - 2a) + (2a - a^2) = 0$$

Answer: (A) $2a - a^2$

38. A circle of radius 5 is circumscribed about a triangle. Another circle of radius 3 is inscribed inside the triangle. Find the area of the triangle.

(A) 24 (B) 32 (C) 12 (D) 16 (E) 17

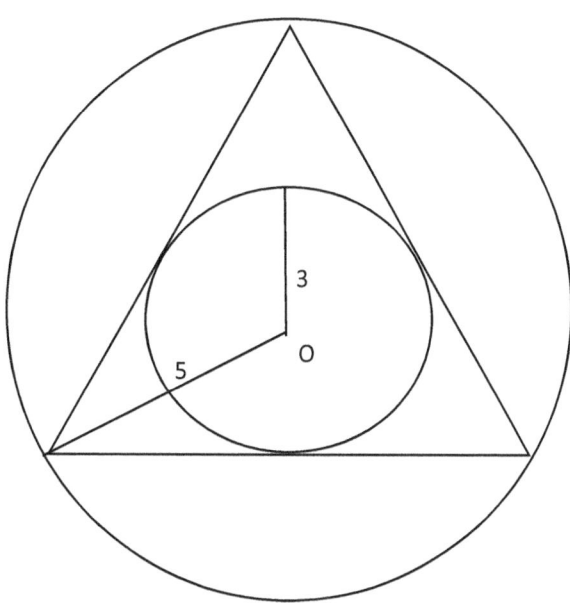

Draw a radius of the smaller circle to the base of the triangle. Find the height of the triangle by adding the two radii. Then use the Pythagorean Theorem to find the base. Finally, use the formula $A = \frac{1}{2}bh$ to find the area of the triangle.

38. A circle of radius 5 is circumscribed about a triangle. Another circle of radius 3 is inscribed inside the triangle. Find the area of the triangle.

(A) 24 (B) 32 (C) 12 (D) 16 (E) 17

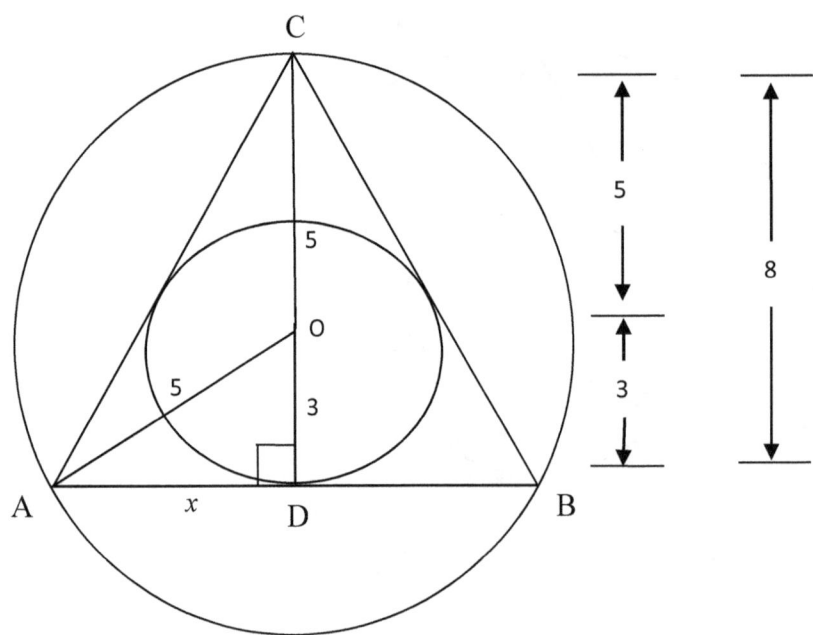

If we extend the radius of the smaller circle perpendicular to base AB, we have right triangle AOD and we can apply the Pythagorean Theorem:

$$a^2 + b^2 = c^2$$
$$3^2 + x^2 = 5^2$$
$$9 + x^2 = 25$$
$$x^2 = 16$$
$$x = 4$$

The height of larger triangle ABC equals the sum of the radii of the two circles, $3 + 5 = 8$.

The base equals $2x = (2)(4) = 8$.

Area of triangle $= \frac{1}{2}bh$, where $b = 2x = 8$.

$$= \frac{1}{2}(8)(8)$$
$$= 32$$

Answer: (B) 32

39. In a survey of local residents, 1/5 were left-handed and 1/12 were ambidextrous. The rest were right-handed. What is the ratio of right-handed people to everyone else?

(A) 29/37   (B) 34/47   (C) 39/65   (D) 43/17   (E) 40/19

Add the fractions representing left-handed residents to the fraction representing ambidextrous residents. Subtract from 1. This represents the fraction of right-handed people. Now develop a ratio of right-handed people to the fraction representing the total of left-handed and ambidextrous people.

39. In a survey of local residents, 1/5 were left-handed and 1/12 were ambidextrous. The rest were right-handed. What is the ratio of right-handed people to everyone else?

(A) 29/37   (B) 34/47   (C) 39/65   (D) 43/17   (E) 40/19

left-handed: $\dfrac{1}{5} = \dfrac{12}{60}$

(+) ambidextrous: $\dfrac{1}{12} = \dfrac{5}{60}$

$\dfrac{17}{60}$

Total: $\dfrac{60}{60}$ (or 1)

(−) left and ambidextrous: $\dfrac{17}{60}$

right-handed: $\dfrac{43}{60}$

$\dfrac{\text{right-handed}}{\text{everyone else}} = \dfrac{43/60}{17/60} = \dfrac{43}{17}$

Answer: (D) 43/17

40. Write the equation of the line which passes through points (2, 5) and (−1, −4).

(A) $y = 2x - 2$ (B) $y = 3x - 1$ (C) $y = 2x + 2$ (D) $y = 3x + 2$ (E) $y = 2x + 1$

Use the formula $y = mx + b$, where m = slope and b = y-intercept. Substitute the co-ordinates of the first point into this general linear equation. Then substitute the co-ordinates of the second point into a second linear equation. You now have two linear equations that can be manipulated in order to solve for the two variables.

41. What happens to the graph of the function $y = x^2 - 2x - 8$ when it's changed to $y = x^2 - 2x - 5$?

(A) flips over   (B) opens to the right   (C) moves down
(D) moves up   (E) opens to the left

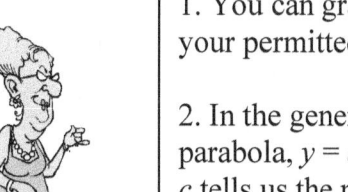

1. You can graph the two functions on your permitted calculator.

2. In the general equation for a parabola, $y = ax^2 + bx + c$, the constant c tells us the number of units to move up (+) or down (−).

40. Write the equation of the line which passes through points (2, 5) and (–1, –4).

(A) $y = 2x - 2$ (B) $y = 3x - 1$ (C) $y = 2x + 2$ (D) $y = 3x + 2$ (E) $y = 2x + 1$

$y = mx + b$, where m = slope and b = y-intercept

(2, 5): $5 = m(2) + b$ (1)
            $2m + b = 5$

(–1, –4): $-4 = m(-1) + b$
            $-1m + b = -4$ (2)

(1) $2m + b = 5$
(2) (–) $-1m + b = -4$
            $3m = 9$
            $m = 3$

(1) $2m + b = 5$
$m = 3$: $2(3) + b = 5$
            $6 + b = 5$
            $b = -1$

$y = mx + b$
$y = 3x - 1$

Answer: (B) $y = 3x - 1$

41. What happens to the graph of the function $y = x^2 - 2x - 8$ when it's changed to $y = x^2 - 2x - 5$?

(A) flips over   (B) opens to the right   (C) moves down
(D) moves up   (E) opens to the left

In the general quadratic equation for a parabola, $y = ax^2 + bx + c$, if the sign preceding the constant c is negative, the parabola moves down c units. In the first equation, $y = x^2 - 2x - 8$, the parabola moves down 8 units. In the second equation, $y = x^2 - 2x - 5$, the parabola moves down only 5 units. So, in comparison, the parabola moves up 3 units.

Answer: (D) moves up

42. Check the circle shown, where O represents the center of the circle, and then select the statement that is definitely correct.

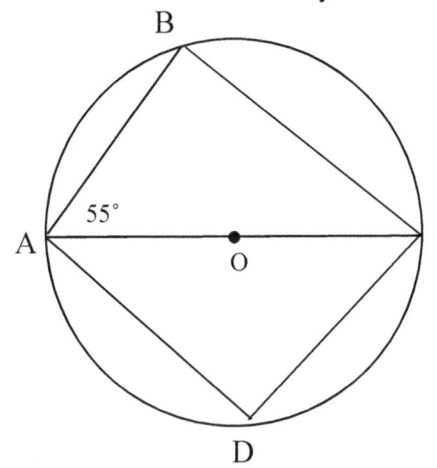

(A) $\widehat{AB} > \widehat{BC}$

(B) $\widehat{BC} > \widehat{AB}$

(C) $\overline{AD} = \overline{CD}$

(D) $\widehat{AB} < \widehat{DC}$

(E) $\widehat{AD} > \widehat{CD}$

Find the missing angles and arcs and then compare.

43. If $f(x) = 2x - 3$ and $g(x) = x^2 + 1$, find $(f \circ g)(3)$.

(A) 15 (B) 13 (C) 17 (D) 12 (E) 19

Find $g(3)$ and substitute the answer into $f(x)$.

42. Check the circle shown, where O represents the center of the circle, and then select the statement that is definitely correct.

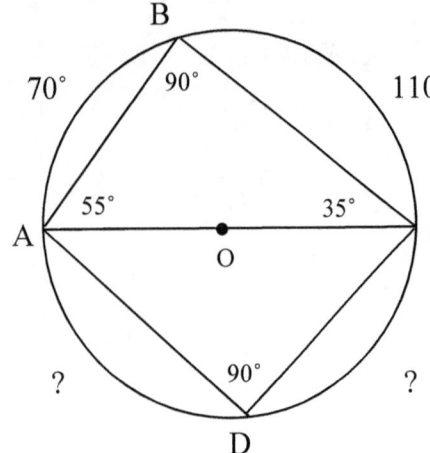

(A) $\overset{\frown}{AB} > \overset{\frown}{BC}$

(B) $\overset{\frown}{BC} > \overset{\frown}{AB}$

(C) $\overline{AD} = \overline{CD}$

(D) $\overset{\frown}{AB} < \overset{\frown}{DC}$

(E) $\overset{\frown}{AD} > \overset{\frown}{CD}$

(A) $\overset{\frown}{AB} > \overset{\frown}{BC}$
    70° > 110°   ✗

(B) $\overset{\frown}{BC} > \overset{\frown}{AB}$
    110° > 70°   ✓

(C) $\overline{AD} = \overline{CD}$?

(D) $\overset{\frown}{AB} < \overset{\frown}{DC}$
    70° < DC ?

(E) $\overset{\frown}{AD} > \overset{\frown}{CD}$?

Answer: (B) $\overset{\frown}{BC} > \overset{\frown}{AB}$

43. If $f(x) = 2x - 3$ and $g(x) = x^2 + 1$, find $(f \circ g)(3)$.

(A) 15   (B) 13   (C) 17   (D) 12   (E) 19

First, find $g(3)$:

$$g(3) = 3^2 + 1 = 10$$

Now, find $f(10)$:

$$f(10) = 2(10) - 3 = 17$$

Answer: (C) 17

44. The probability of rain is .1. The probability of sun is .8, and the probability of a sun shower is .02. What is the probability of either rain or sun but not both?

(A) .76 (B) .91 (C) .88 (D) .96 (E) .81

When two probabilities are not mutually exclusive,

P(A or B) = P(A) + P(B) − P(A and B)

45. What is the equation of the graph shown?

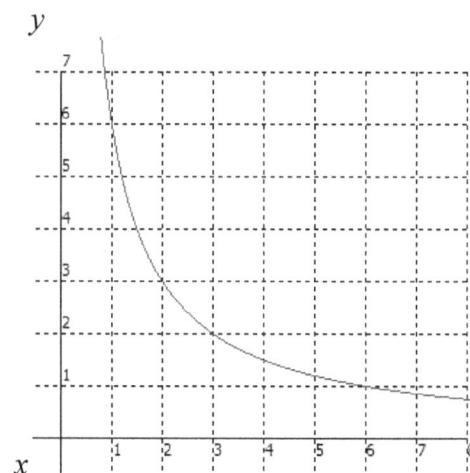

(A) $xy = 12$
(B) $y = 4$
(C) $xy = 6$
(D) $x^2 + y^2 = 9$
(E) $y = 2x^2 + 4$

We can first eliminate the linear equations. Then locate 3 points along the graph and see which of the remaining equations fit the graph.

44. The probability of rain is .1. The probability of sun is .8, and the probability of a sun shower is .02. What is the probability of either rain or sun but not both?

(A) .76   (B) .91   (C) .88   (D) .96   (E) .81

Let A = probability of rain = .1
Let B = probability of sun = .8.
The probability of a sun shower = .02

$$P(A \text{ or } B) = P(A) + P(B) - P(A \text{ and } B) =$$
$$.1 + .8 - .02 =$$
$$.88$$

Answer: (C) .88

45. What is the equation of this graph?   (1, 6)

(A) $xy = 12$
(B) $y = 4$
(C) $xy = 6$
(D) $x^2 + y^2 = 9$
(E) $y = 2x^2 + 4$

We can immediately eliminate (B), because it's a linear equation.

Selection (D) is the equation of a circle, so we'll drop it.

Selection (E) is the equation of a parabola, so it isn't the answer.

Let's test (A), $xy = 12$.

(6, 1): $(6)(1) \neq 12$

Now, let's try (C), $xy = 6$.

(6, 1): $(6)(1) = 6$
(3, 2): $(3)(2) = 6$
(2, 3): $(2)(3) = 6$

Answer: (C) $xy = 6$

46. The volume of a pump varies inversely with the pressure applied. If a pressure of 20 lbs. is applied, the volume in the pump reads 100 cubic inches. What is the volume if 25 lbs. of pressure is applied?

(A) 80 cu in   (B) 60 cu in   (C) 100 cu in   (D) 90 cu in   (E) 70 cu in

In inverse variation when one factor increases, the other factor decreases. Their product is a constant.

47. In a sample population, *a* respondents are each *x* years old, *b* respondents are each *y* years old and *c* respondents are each *z* years old. What is their average age?

(A) $\dfrac{a+b+c}{x+y+z}$ (B) $\dfrac{abc+xyz}{ax+yb+cz}$ (C) $\dfrac{ab+bc+abc}{abc}$ (D) $\dfrac{ax+by+cz}{a+b+c}$

(E) $\dfrac{ab+b+ac}{xy+z}$

Multiply the number of respondents by their age, and then add up the ages and divide by the total number of respondents.

46. The volume of a pump varies inversely with the pressure applied. If a pressure of 20 lbs. is applied, the volume in the pump reads 100 cubic inches. What is the volume if 25 lbs. of pressure is applied?

(A) 80 cu in   (B) 60 cu in   (C) 100 cu in   (D) 90 cu in   (E) 70 cu in

Let P_1 and P_2 equal the two pressures.
Let V_1 and V_2 equal the corresponding two volumes.
k is the constant.

$P_1 = 20, V_1 = 100$:
$$P_1 V_1 = k$$
$$20(100) = k$$
$$k = 2000$$

$P_2 = 25, k = 2000$:
$$P_2 V_2 = k$$
$$25 V_2 = 2000$$
$$V_2 = 80$$

Answer: (A) 80 cu in

47. In a sample population, a respondents are each x years old, b respondents are each y years old and c respondents are each z years old. What is their average age?

(A) $\dfrac{a+b+c}{x+y+z}$ (B) $\dfrac{abc+xyz}{ax+yb+cz}$ (C) $\dfrac{ab+bc+abc}{abc}$ (D) $\dfrac{ax+by+cz}{a+b+c}$

(E) $\dfrac{ab+b+ac}{xy+z}$

If a respondents are each x years old, then ax is the sum of the ages of a respondents. Similarly, by and cz are the sums of the ages of the b and c respondents, respectively.

The total number of respondents is $a + b + c$, so the fraction is

$$\frac{ax+by+cz}{a+b+c}$$

Answer (D) $\dfrac{ax+by+cz}{a+b+c}$

48. The chart below indicates the inches of rain that fell per year and the number of times each yearly accumulation occurred for the past 15 years. What was the mean number of inches of rain that fell per year for that period of time?

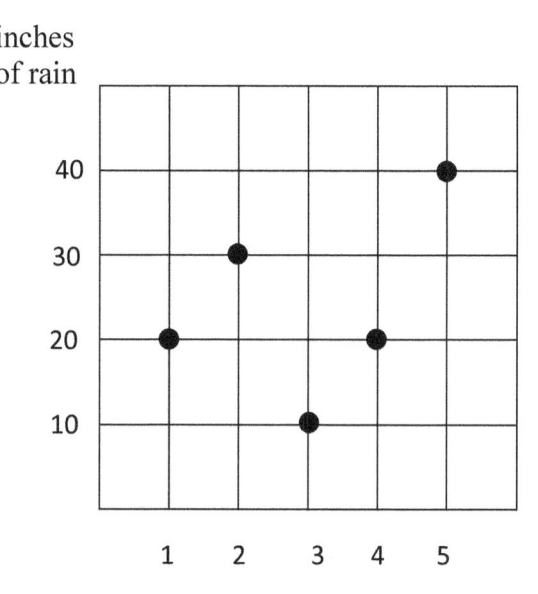

Multiply the number of years by the number of inches of rain, add them all and divide by the total number of years.

(A) 26
(B) 19
(C) 32
(D) 35
(E) 24

49. In the rectangular solid shown, FG = 3, EF = 4 and BG = 6. Find diagonal BE.

(A) $\sqrt{56}$ (B) $\sqrt{61}$ (C) $2\sqrt{39}$ (D) $\sqrt{76}$ (E) $3\sqrt{17}$

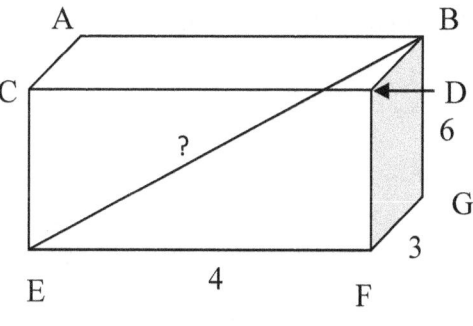

Draw EG and use the Pythagorean Theorem.

48. The chart below indicates the inches of rain that fell per year and the number of times each yearly accumulation occurred for the past 15 years. What was the mean number of inches of rain that fell per year for that period of time?

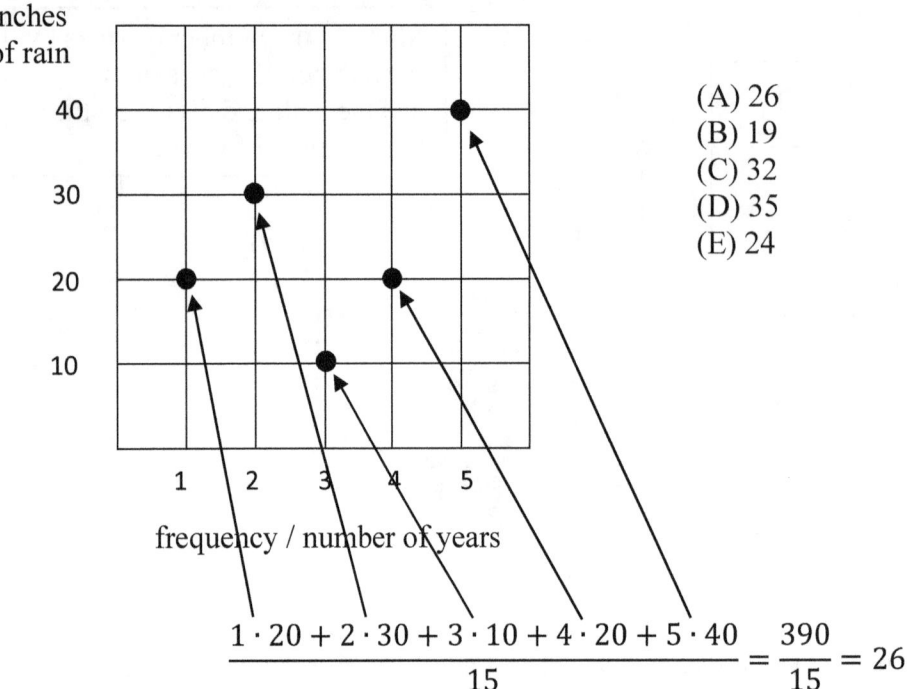

(A) 26
(B) 19
(C) 32
(D) 35
(E) 24

$$\frac{1 \cdot 20 + 2 \cdot 30 + 3 \cdot 10 + 4 \cdot 20 + 5 \cdot 40}{15} = \frac{390}{15} = 26$$

Answer: (E) 24

49. In the rectangular solid shown, FG = 3, EF = 4 and BG = 6. Find diagonal BE.

(A) $\sqrt{56}$ (B) $\sqrt{61}$ (C) $2\sqrt{39}$ (D) $\sqrt{76}$ (E) $3\sqrt{17}$

Draw EG, so that we have right $\triangle$ EFG in the base of the rectangular solid.

Now we can apply the Pythagorean Theorem: $3^2 + 4^2 = EG^2$
$9 + 16 = EG^2$
$EG = 5$

We have another right triangle, $\triangle EGB$.

Once again, we can use the Pythagorean Theorem:
$5^2 + 6^2 = BE^2$
$25 + 36 = BE^2$
$\sqrt{61} = BE$

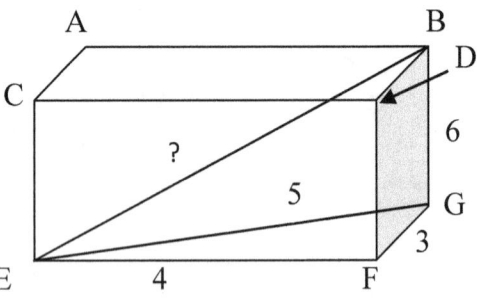

Answer: (B) $\sqrt{61}$

50. Which of the following illustrations represents a function?

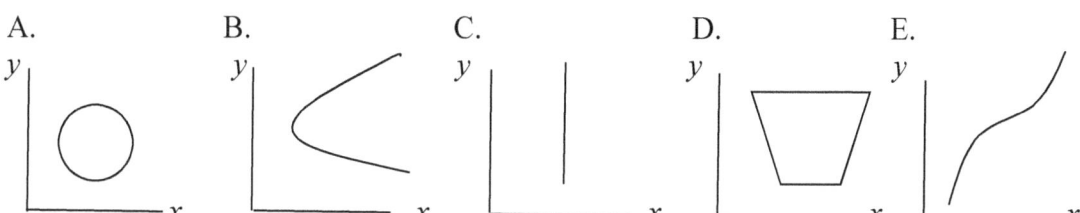

A.   B.   C.   D.   E.

In a function, each element in the domain, *x*, is mapped into only one element in the range, *y*.

51. If "*a*" represents a true statement and "*b*" represents a false statement, determine which column is true?

a	b	column 1	column 2	column 3	column 4	column 5
T	F	$a \wedge b$	$\sim a \vee b$	$\sim b \wedge \sim a$	$b \vee (b \wedge \sim a)$	$a \vee (b \wedge \sim a)$

(A) column 1   (B) column 4   (C) column 3   (D) column 2   (E) column 5

$\sim$ = not
$\vee$ = or
$\wedge$ = and

For an "and" ($\wedge$) statement to be true, both parts of the statement have to be true. For an "or" ($\vee$) statement, only one part of the statement has to be true. The negation ($\sim$) of a statement returns the reverse truth value of the original statement.

50. Which of the following illustrations represents a function?

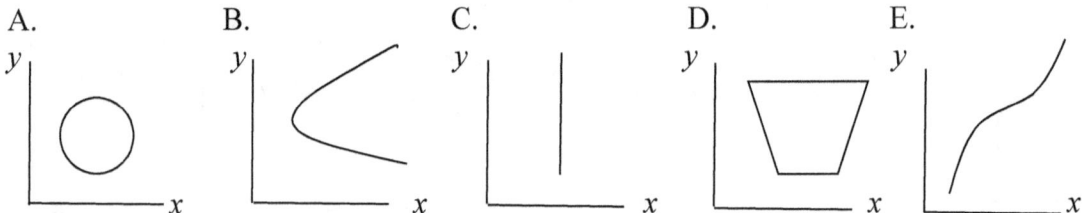

In a function each element in the domain, x, is mapped into only one element in the range, y.

In cases A, B and D, one element in the domain (x-value) is mapped into two elements in the range (y-value). So, these are not functions. In case C, one x is mapped into many y's, so C is not a function.

However, in case E each of the elements in the domain is mapped into only one element in the range. You may recall the vertical line test for a function: Each vertical must intersect the graph at most one time.

Answer: (E)

51. If "a" represents a true statement and "b" represents a false statement, determine which column is true?

a	b	column 1	column 2	column 3	column 4	column 5
T	F	a ∧ b	~a ∨ b	~b ∧ ~a	b ∨ (b ∧ ~a)	a ∨ (b ∧ ~a)

(A) column 1   (B) column 4   (C) column 3   (D) column 2   (E) column 5

~ = not
∨ = or
∧ = and

		column 1	column 2	column 3	column 4	column 5
a	b	a ∧ b	~a ∨ b	~b ∧ ~a	b ∨ (b ∧ ~a)	a ∨ (b ∧ ~a)
T	F	T ∧ F	F ∨ F	T ∧ F	F ∨ (F ∧ F)	T ∨ (F ∧ F)
					F ∨ F	T ∨ F
		F	F	F	F	T

Answer: (E) column 5

52. Write the quadratic equation one of whose roots is $4 + 3i$.

(A) $x^2 - 8x + 25 = 0$ (B) $x^2 - 2x + 8 = 0$ (C) $x^2 - 2x - 8 = 0$ (D) $x^2 - 6x + 21 = 0$
(E) $x^2 - 3x + 6 = 0$

The other root is the conjugate $4 - 3i$.

53. If $a : b = c : d$ and $d = a/2$, find the value of b.

(A) $\dfrac{3c}{2a}$ (B) $2ac^2$ (C) $\dfrac{3a^2}{c}$ (D) $\dfrac{a^2}{2c}$ (E) $\dfrac{3a}{2c^2}$

Substitute $a/2$ for d in the proportion.

52. Write the quadratic equation one of whose roots is $4 + 3i$.

(A) $x^2 - 8x + 25 = 0$ (B) $x^2 - 2x + 8 = 0$ (C) $x^2 - 2x - 8 = 0$ (D) $x^2 - 6x + 21 = 0$
(E) $x^2 - 3x + 6 = 0$

If one root is $4 + 3i$, its conjugate, $4 - 3i$, is also a root. Multiply to get the original quadratic expression.

$$\begin{array}{r} x - 4 + 3i \\ \underline{x - 4 - 3i} \\ -3ix + 12i - 9i^2 \\ -4x + 16 - 12i \\ \underline{x^2 - 4x + 3ix } \\ x^2 - 8x + 16 + 9 \\ x^2 - 8x + 25 \end{array}$$

Answer: (A) $x^2 - 8x + 25 = 0$

53. If $a : b = c : d$ and $d = a/2$, find the value of b.

(A) $\dfrac{3c}{2a}$ (B) $2ac^2$ (C) $\dfrac{3a^2}{c}$ (D) $\dfrac{a^2}{2c}$ (E) $\dfrac{3a}{2c^2}$

Writing the ratios as fractions yields

$$\frac{a}{b} = \frac{c}{d}$$

Cross-multiplying yields

$$ad = bc$$

Replacing d with $a/2$ yields

$$a \cdot \frac{a}{2} = bc$$

$$\frac{a^2}{2} = bc$$

$$b = \frac{a^2}{2c}$$

Answer: (D) $b = \dfrac{a^2}{2c}$

54. What is the range of the function $y = 2^x$?

(A) $-\infty < x < \infty$ (B) $y \geq 0$ (C) $y > 0$ (D) $0 < x$ (E) $y \geq 0$

Solve by either of the following ways:

1. Plot on your graphing calculator.

2. Simplify the problem by setting $-\infty < x < +\infty$ and view the range, the y-value.

55. Within the set of real numbers, what is the domain of the following function:

$$f(x) = \frac{5x + 2}{\sqrt{x^2 - 5}}$$

(A) $|x| > \sqrt{5}$ (B) $-5 < x < 5$ (C) $x < 5$ (D) $x < \sqrt{5}$ (E) $\sqrt{5} < x < 5$

The denominator cannot equal zero because that would cause division by 0, and the denominator cannot be negative because that would create a complex number.

54. What is the range of the function $y = 2^x$?

(A) $-\infty < x < \infty$ (B) $y \geq 0$ (C) $y > 0$ (D) $0 < x$ (E) $y \geq 0$

Let's either plot it on a graphing calculator or plot it manually.

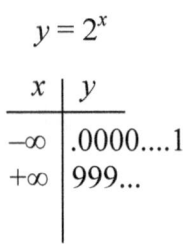

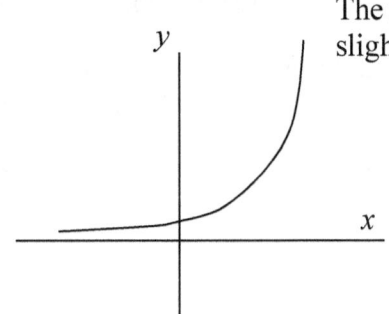

The range of the y-values is slightly above 0 to ∞.

Answer: (C) $y > 0$

55. Within the set of real numbers, what is the domain of the following function:

$$f(x) = \frac{5x + 2}{\sqrt{x^2 - 5}}$$

(A) $|x| > \sqrt{5}$ (B) $-5 < x < 5$ (C) $x < 5$ (D) $x < \sqrt{5}$ (E) $\sqrt{5} < x < 5$

$x^2 - 5 \neq 0$, which makes the fraction undefined (you cannot divide by zero).

On the other hand, $x^2 - 5$ cannot be negative because that would create a complex number. So, $x^2 - 5$ must be strictly greater than zero:

$$x^2 - 5 > 0$$

$$x^2 > 5$$

$$\sqrt{x^2} > \sqrt{5}$$

Recall that one of the definitions of absolute value is $|x| = \sqrt{x^2}$. Hence, we get

$$|x| > \sqrt{5}$$

Answer: (A) $|x| > \sqrt{5}$

56. Change $1\frac{2}{3}\pi$ radians to degrees.

(A) 180°   (B) 210°   (C) 270°   (D) 300°   (E) 340°

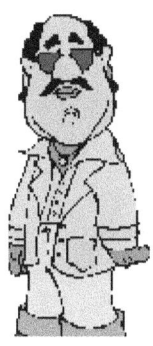

2π radians = 360°. Now, set up a proportion.

57. Manuel charges 5% for his financial advice. If a client had $7,600 left after Manuel was paid, how much did the client have originally?

(A) $9,000   (B) $9,500   (C) $8,000   (D) $10,000   (E) $10,500

Set the percentage of the investment left after Manuel's commission equal to the final total.

56. Change $1\frac{2}{3}\pi$ radians to degrees.

(A) 180°   (B) 210°   (C) 270°   (D) 300°   (E) 340°

Set up a proportion. Let x be the equivalent degree measure of $1\frac{2}{3}\pi$. Then the ratio of radians is proportional to the ratio of degrees:

$$\frac{1\frac{2}{3}\pi}{2\pi} = \frac{x}{360}$$

$$\frac{5\pi/3}{2\pi} = \frac{x}{360}$$

$$\frac{5}{6} = \frac{x}{360}$$

$$\frac{5}{6} \cdot 360 = x$$

$$x = 300$$

Answer: (D) 300°

57. Manuel charges 5% for his financial advice. If a client had $7,600 left after Manuel was paid, how much did the client have originally?

(A) $9,000   (B) $9,500   (C) $8,000   (D) $10,000   (E) $10,500

Let x be the money in the original account.

$$x - .05x = 7,600$$
$$.95x = 7,600$$
$$x = 8,000$$

Answer: (C) $8,000

58. What is the period of the function $y = 2 \sin 3x$?

(A) $2\pi/3$ (B) $\pi/2$ (C) 2π (D) π (E) $3\pi/4$

In the general form of the function $y = a \sin bx + k$, the period is equal to $2\pi/b$.

59. Find the point of symmetry (where the diagonals intersect) of parallelogram ABCD.

(A) $(4\frac{1}{2}, 6)$ (B) $(3, 5)$ (C) $(3\frac{1}{2}, 4\frac{1}{2})$ (D) $(3\frac{1}{2}, 5)$ (E) $(3\frac{1}{2}, 4)$

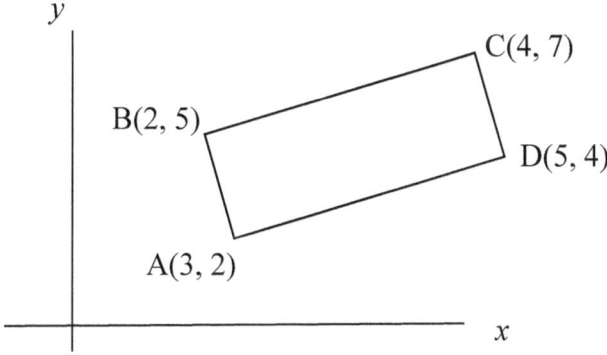

Draw diagonals AC and BD and find the common midpoint.

58. What is the period of the function $y = 2 \sin 3x$?

(A) $2\pi/3$ (B) $\pi/2$ (C) 2π (D) π (E) $3\pi/4$

A periodic function is one that repeats itself, so we have to determine how often $y = 2 \sin 3x$ repeats itself.

In the general form of the function $y = a \sin bx + k$, the period is equal to $2\pi/b$.

In the given function ($y = 2 \sin 3x$), 3 is in the position of b in the general form. So, $b = 3$ and $2\pi/b$ becomes $2\pi/3$.

Answer: (A) $2\pi/3$

59. Find the point of symmetry (where the diagonals intersect) of parallelogram ABCD.

(A) $(4\frac{1}{2}, 6)$ (B) $(3, 5)$ (C) $(3\frac{1}{2}, 4\frac{1}{2})$ (D) $(3\frac{1}{2}, 5)$ (E) $(3\frac{1}{2}, 4)$

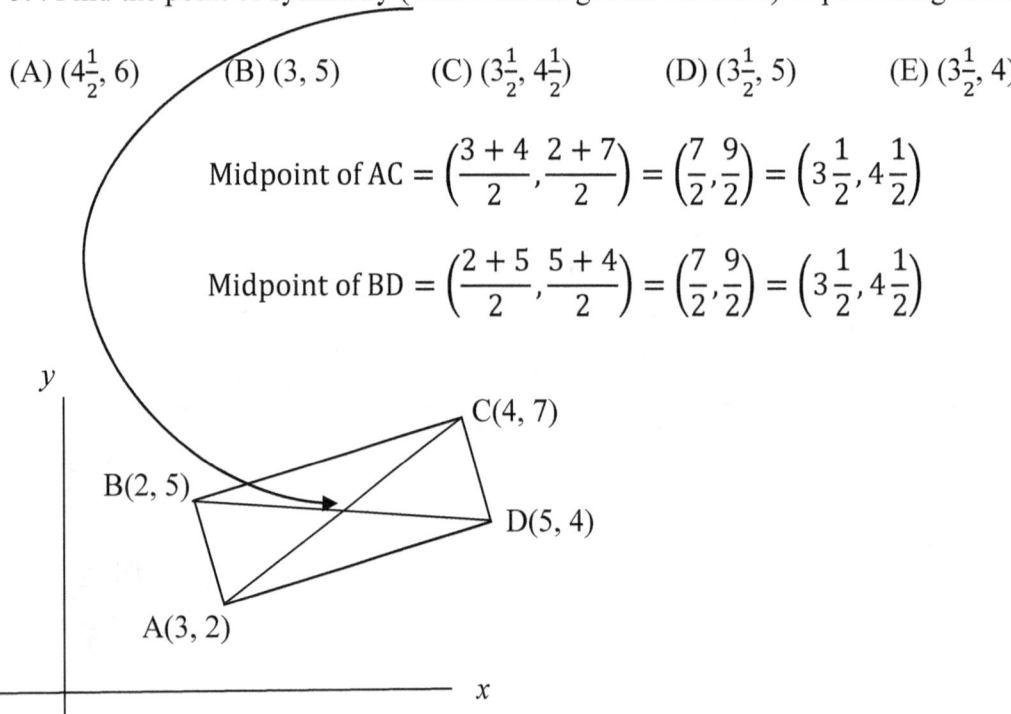

$$\text{Midpoint of AC} = \left(\frac{3+4}{2}, \frac{2+7}{2}\right) = \left(\frac{7}{2}, \frac{9}{2}\right) = \left(3\frac{1}{2}, 4\frac{1}{2}\right)$$

$$\text{Midpoint of BD} = \left(\frac{2+5}{2}, \frac{5+4}{2}\right) = \left(\frac{7}{2}, \frac{9}{2}\right) = \left(3\frac{1}{2}, 4\frac{1}{2}\right)$$

Answer: (C) $(3\frac{1}{2}, 4\frac{1}{2})$

60. What is the sum of the first 16 terms of the arithmetic sequence if $a_3 = -14$ and $a_4 = -18$.

(A) –384 (B) –492 (C) –678 (D) –488 (E) –576

To find the sum of an Arithmetic Sequence, use the formula
$$S_n = \frac{n}{2}(a_1 + a_n)$$
where S_n = the sum of n terms, n = the number of terms, a_1 = the first term and a_n = the last term. What we're actually doing is getting the average of the first and last terms and multiplying by the number of terms. You'll also need the formula $a_n = a_1 + (n-1)d$, where d = the common difference, in order to find the last term.

60. What is the sum of the first 16 terms of the arithmetic sequence if $a_3 = -14$ and $a_4 = -18$.

(A) –384 (B) –492 (C) –678 (D) –488 (E) –576

Let's use the formula $S_n = n(a_1 + a_n)/2$, where a_1 = the first term, n = the number of terms, a_n = the last term, and d = the common difference.

If $a_4 = -18$ and $a_3 = -14$, then $d = a_4 - a_3 = -18 - (-14) = -18 + 14 = -4$.

To use the formula $S_n = n(a_1 + a_n)/2$, we need to find the first term, a_1. We'll use the formula $a_n = a_1 + (n-1)d$. Let's take a_4 as the nth term.

$$a_4 = a_1 + (4-1)(-4)$$
$$-18 = a_1 + 3(-4)$$
$$-18 = a_1 - 12$$
$$a_1 = -6$$

Now, let's find the last term, a_n, the 16th term.

$a_n = a_1 + (n-1)d$
$a_{16} = -6 + (16-1)(-4)$
$a_{16} = -6 + (15)(-4)$
$a_{16} = -6 - 60 = -66$

We'll now go back to use the formula:

$$S_n = \frac{n}{2}(a_1 + a_n) = \frac{16}{2}(-6 + [-66]) = 8(-72) = -576$$

Answer: (E) –576

Answer Key to Test 1

1. D	21. E	41. D
2. B	22. A	42. B
3. E	23. A	43. C
4. B	24. A	44. C
5. C	25. E	45. C
6 E	26. E	46. A
7. C	27. C	47. D
8. D	28. A	48. E
9. A	29. B	49. B
10. E	30. A	50. E
11. B	31. E	51. E
12. D	32. A	52. A
13. E	33. E	53. D
14. A	34. B	54. C
15. C	35. E	55. A
16. B	36. B	56. D
17. D	37. A	57. C
18. B	38. B	58. A
19. D	39. D	59. C
20. D	40. B	60. E

Test 2

1. A truck tire has a radius of .25 meters. If the tire rotates at the rate of 400 revolutions per minute, how long will it take to cover a distance of 5 kilometers? For this problem, let π = 3.14 and round off your answer to the nearest minute. One kilometer equals 1,000 meters. Round off to the nearest minute.

(A) 4 minutes   (B) 8 minutes   (C) 7 minutes   (D) 9 minutes   (E) 10 minutes

First, find the circumference of the tire. Multiply by 400 to find the distance covered in one minute. Finally, divide 5 kilometers (5,000 meters) by the distance covered in one minute.

2. Jason has $900 to spend. If he buys a laptop at a 20% discount from the original price of $520, a cell phone at a discount of 30% from the original price of $200 and a tablet at a discount of 40% from the original price of $500, how much money does he have left?

(A) $26   (B) $53   (C) $49   (D) $38   (E) $44

After a discount, find the price Jason pays for each item, add them together and subtract from $900.

1. A truck tire has a radius of .25 meters. If the tire rotates at the rate of 400 revolutions per minute, how long will it take to cover a distance of 5 kilometers? For this problem, let $\pi = 3.14$ and round off your answer to the nearest minute. One kilometer equals 1,000 meters. Round off to the nearest minute.

(A) 4 minutes (B) 8 minutes (C) 7 minutes (D) 9 minutes (E) 10 minutes

1 kilometer = 1,000 meters
5 kilometers = 5,000 meters
$\pi = 3.14$

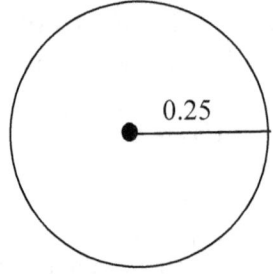

Circumference of the tire: $2\pi \cdot \text{radius} = 2\pi(.25\text{m}) = .50\pi$ meters
Distance travelled in 400 revolutions: $.50\pi \times 400 = 200\pi$ meters in one minute

Recall that $Distance = Rate \cdot Time$ ($D = R \cdot T$). Since the question is asking for time, solve this equation for T:

$$T = \frac{D}{R}$$

$$\text{Time} = \frac{5000 \text{ meters}}{200\pi \text{ meters per minute}} = \frac{5000}{200 \times 3.14} = \frac{5000}{628} = 7.96\ldots \approx 8 \text{ minutes}$$

Answer: (B) 8 minutes

2. Jason has $900 to spend. If he buys a laptop at a 20% discount from the original price of $520, a cell phone at a discount of 30% from the original price of $200 and a tablet at a discount of 40% from the original price of $500, how much money does he have left?

(A) $26 (B) $53 (C) $49 (D) $38 (E) $44

For a laptop, Jason pays .80 × $520 = $416
For a cell phone, Jason pays .70 × $200 = $140
For a tablet, Jason pays .60 × $500 = $300
                              Total = $856

```
      $900
(−)   $856
       $44
```

Answer: (E) $44

3. Simplify $\dfrac{n!}{(n-x)!}$, where n and x are positive integers, $n > x$ and $x = 3$.

(A) $n(n-1)(n-2)(n-3)$ (B) $n(n-1)(n-2)$ (C) $n(n-1)$
(D) $n(n-1)(n-2)(n-3)(n-4)$ (E) $n(n-2)(n-4)$

$n! = n(n-1)(n-2)\ldots(1)$

4. The ratio of $a : b$ is $3 : 5$. The ratio of $c : d$ is $4 : 5$. Find the ratio of $a : c$.

(A) $\dfrac{2b}{3d}$ (B) $\dfrac{5}{3d}$ (C) $\dfrac{4b}{5}$ (D) $\dfrac{3b}{4d}$ (E) $\dfrac{3b}{5d}$

Set up two equations. Solve for a and c and then form their ratio.

3. Simplify $\dfrac{n!}{(n-x)!}$, where n and x are positive integers, $n > x$ and $x = 3$.

(A) $n(n-1)(n-2)(n-3)$ (B) $n(n-1)(n-2)$ (C) $n(n-1)$
(D) $n(n-1)(n-2)(n-3)(n-4)$ (E) $n(n-2)(n-4)$

$$\frac{n!}{(n-3)!} = \frac{n(n-1)(n-2)\cancel{(n-3)(n-4)}\ldots \cancel{1}}{\cancel{(n-3)(n-4)(n-5)(n-6)}\ldots \cancel{1}} = n(n-1)(n-2)$$

Answer: (B) $n(n-1)(n-2)$

4. The ratio of $a : b$ is $3 : 5$. The ratio of $c : d$ is $4 : 5$. Find the ratio of $a : c$.

(A) $\dfrac{2b}{3d}$ (B) $\dfrac{5}{3d}$ (C) $\dfrac{4b}{5}$ (D) $\dfrac{3b}{4d}$ (E) $\dfrac{3b}{5d}$

Forming the proportion $a : b$ as $3 : 5$ yields

$$\frac{a}{b} = \frac{3}{5}$$

Forming the proportion $c : d$ as $4 : 5$ yields

$$\frac{c}{d} = \frac{4}{5}$$

Solving for a and c (since we are looking for the ratio of $a : c$) yields

$$a = \frac{3b}{5} \qquad c = \frac{4d}{5}$$

Forming the ratio $a : c$ yields

$$\frac{a}{c} = \frac{3b}{5} \div \frac{4d}{5} = \frac{3b}{5} \times \frac{5}{4d} = \frac{3b}{4d}$$

Answer: (D) $\dfrac{3b}{4d}$

5. Find the value of n in the equation

$$\left[\frac{3}{5}\right]^n + \left[\frac{2}{5}\right]^2 = \frac{47}{125}$$

(A) 5 (B) 4 (C) 3 (D) 2 (E) 6

Simplify $\left[\frac{2}{5}\right]^2$ and then solve for n.

6.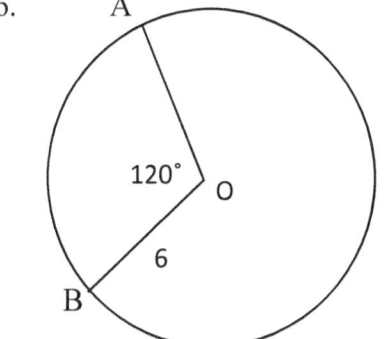

In the circle O shown, radius OB = 6, and central angle AOB = 120°. Find the measure of minor arc AB.

(A) 4π (B) 12π (C) 8π (D) 10π (E) 6π

Find the circumference of the circle. Then determine what fraction of that circumference the minor arc AB occupies.

5. Find the value of n in the equation

$$\left[\frac{3}{5}\right]^n + \left[\frac{2}{5}\right]^2 = \frac{47}{125}$$

(A) 5   (B) 4   (C) 3   (D) 2   (E) 6

$$\left[\frac{3}{5}\right]^n + \frac{4}{25} = \frac{47}{125}$$

$$\left[\frac{3}{5}\right]^n = \frac{47}{125} - \frac{4}{25} = \frac{47}{125} - \frac{20}{125} = \frac{27}{125} = \frac{3^3}{5^3} = \left[\frac{3}{5}\right]^3$$

Since the bases (3/5) are equal, the exponents must be equal:

$$n = 3$$

Answer: (C) 3

6.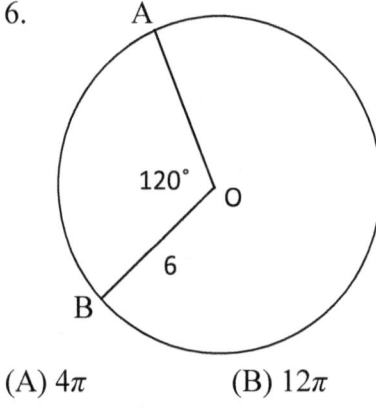

In the circle O shown, radius OB = 6, central angle AOB = 120°. Find the measure of minor arc AB.

(A) 4π (B) 12π (C) 8π (D) 10π (E) 6π

The circumference of the circle is $2\pi r = 2\pi(6) = 12\pi$. Now, the central angle AOB is one-third of the circumference:

$$\frac{120°}{360°} = \frac{1}{3}$$

So, the measure of minor arc AB is one-third of the circumference:

$$\frac{1}{3} \cdot 12\pi = 4\pi$$

Answer: (A) 4π

7. Find the value of *b*.

$$(1)\ 2a + 3b = -3$$
$$(2)\ 2b = 4a - 34$$

(A) 3 (B) –5 (C) 4 (D) –6 (E) –2

Line up *a* and *b*. Eliminate either variable and then solve for the remaining one.

8. José is on the track team. He wants to run 2 hours a day. However, he has to work after school and also help his mother with the baby. If he has a maximum of 6 hours for all three activities and he works twice as long as he helps with the baby, what is the greatest amount of time he can work?

(A) $2\frac{1}{3}$ hours (B) $3\frac{1}{2}$ hours (C) $3\frac{1}{3}$ hours (D) $2\frac{1}{2}$ hours (E) $2\frac{2}{3}$ hours

This is an inequality problem. Let *t* = the time with the baby and take everything from there.

7. Find the value of b.

$$(1)\ 2a + 3b = -3$$
$$(2)\ 2b = 4a - 34$$

(A) 3   (B) –5   (C) 4   (D) –6   (E) –2

Multiply (1) by 2: $2(2a + 3b = -3)$
Multiply (2) by 3: $3(-4a + 2b = -34)$

Subtract: (–) $\begin{array}{l} 4a + 6b = -6 \\ -12a + 6b = -102 \end{array}$

$$16a = 96$$
$$a = 6$$

Now, substitute $a = 6$ into equation (1):

$$2(6) + 3b = -3$$
$$12 + 3b = -3$$
$$3b = -15$$
$$b = -5$$

Answer: (B) –5

Note: It would have been faster to eliminate the variable a initially, but we wanted to show the whole process for solving these systems of equations. That is, determine the value of one of the variables and then substitute that value into one of the other equations.

8. José is on the track team. He wants to run 2 hours a day. However, he has to work after school and also help his mother with the baby. If he has a maximum of 6 hours for all three activities and he works twice as long as he helps with the baby, what is the greatest amount of time he can work?

(A) $2\frac{1}{3}$ hours   (B) $3\frac{1}{2}$ hours   (C) $3\frac{1}{3}$ hours   (D) $2\frac{1}{2}$ hours   (E) $2\frac{2}{3}$ hours

Amount of time on the track: 2 hours
Amount of time helping with the baby: t
Amount of time working: $2t$

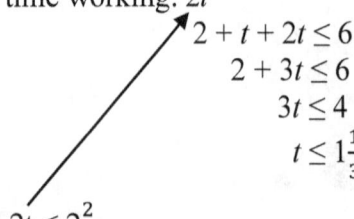

$$2 + t + 2t \leq 6$$
$$2 + 3t \leq 6$$
$$3t \leq 4$$
$$t \leq 1\frac{1}{3}$$

$$2t \leq 2\frac{2}{3}$$

Maximum working time = $2\frac{2}{3}$ hours.

Answer: (E) $2\frac{2}{3}$ hours

9. Which of the following functions has an amplitude of 3 and a period of 4π?

(A) $y = -4 \sin 3x + 2$ (B) $y = 3 \sin 4x - 1$ (C) $y = 4 \sin 4x$ (D) $y = 3 \sin \frac{1}{2}x$
(E) $y = 4 \sin \frac{1}{2}x$

$f(x) = a \sin bx + c$, x is in radians

1. $|a|$ is the amplitude.
2. The period is $\frac{2\pi}{b}$.
3. c (+) is the height above the x-axis, c (−) is the height below the x-axis.

10. The following digits satisfy the Δ operation defined as $\dfrac{x \Delta y}{2} = z$:

(1) $x = 3, y = 4, z = 6$ (2) $x = 3, y = 6, z = 9$ (3) $x = 5, y = 4, z = 10$

Find the definition of the Δ operation and then determine the value of y in the equation $\dfrac{x \Delta y}{2} = z$ when $x = 7$ and $z = 14$.

(A) 6 (B) 5 (C) 4 (D) 3 (E) 7

Using the given values for x, y and z for the Δ operation, see which operations satisfy the result in z.

9. Which of the following functions has an amplitude of 3 and a period of 4π?

(A) $y = -4 \sin 3x + 2$ (B) $y = 3 \sin 4x - 1$ (C) $y = 4 \sin 4x$ (D) $y = 3 \sin \frac{1}{2}x$
(E) $y = 4 \sin \frac{1}{2}x$

$$y = a \sin bx + c$$

1. $|a|$ is the amplitude: $|3| = 3$
2. $\frac{2\pi}{b}$ ($b > 0$) is the period: $\frac{2\pi}{b} = 2\pi \div \frac{1}{2} = 4\pi$

Answer: (D) $y = 3 \sin \frac{1}{2}x$

10. The following digits satisfy the Δ operation defined as $\frac{x \Delta y}{2} = z$:

(1) $x = 3$, $y = 4$, $z = 6$ (2) $x = 3$, $y = 6$, $z = 9$ (3) $x = 5$, $y = 4$, $z = 10$

Find the definition of the Δ operation and then determine the value of y in the equation $x \Delta y = z$ when $x = 7$ and $z = 14$.

(A) 6   (B) 5   (C) 4   (D) 3   (E) 7

We don't know what mathematical operation Δ stands for. So, let's assign some basic operations to Δ to see whether we can discern what it stands for. Suppose Δ stands for addition. Then from (1), we get

$$\frac{x \Delta y}{2} = \frac{3 + 4}{2} = \frac{7}{2}$$

But $7/2 \neq 6$, which is the z-value in Statement (1). Hence, the Δ operation cannot be addition.

Now, suppose Δ stands for multiplication. Then from (1), we get

$$\frac{x \Delta y}{2} = \frac{3 \cdot 4}{2} = \frac{12}{2} = 6$$

Here, we do get the value of z ($= 6$) in Statement (1). So, it appears that the Δ operation stands for multiplication. Let's verify this for statements (2) and (3):

$$\frac{x \Delta y}{2} = \frac{3 \cdot 6}{2} = \frac{18}{2} = 9$$

$$\frac{x \Delta y}{2} = \frac{5 \cdot 4}{2} = \frac{20}{2} = 10$$

So, $\frac{x \Delta y}{2} = \frac{x \cdot y}{2} = z$

$$\frac{7 \cdot y}{2} = 14$$

Answer: (C) 4

$y = 4$

11. ABCD and BECD are parallelograms. Angle ADB measures 72° and angle BDC measures 67°. Find $m\angle A$.

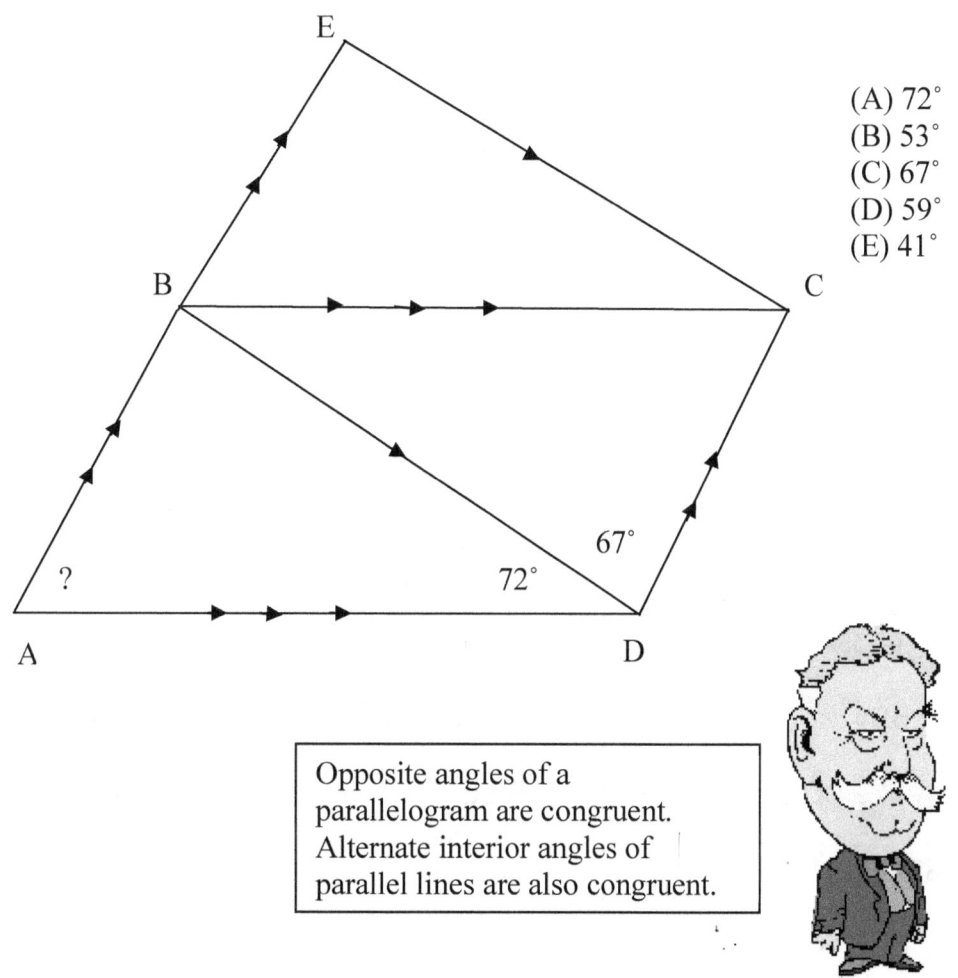

(A) 72°
(B) 53°
(C) 67°
(D) 59°
(E) 41°

Opposite angles of a parallelogram are congruent. Alternate interior angles of parallel lines are also congruent.

11. ABCD and BECD are parallelograms. Angle ADB measures 72° and angle BDC measures 67°. Find $m\angle A$.

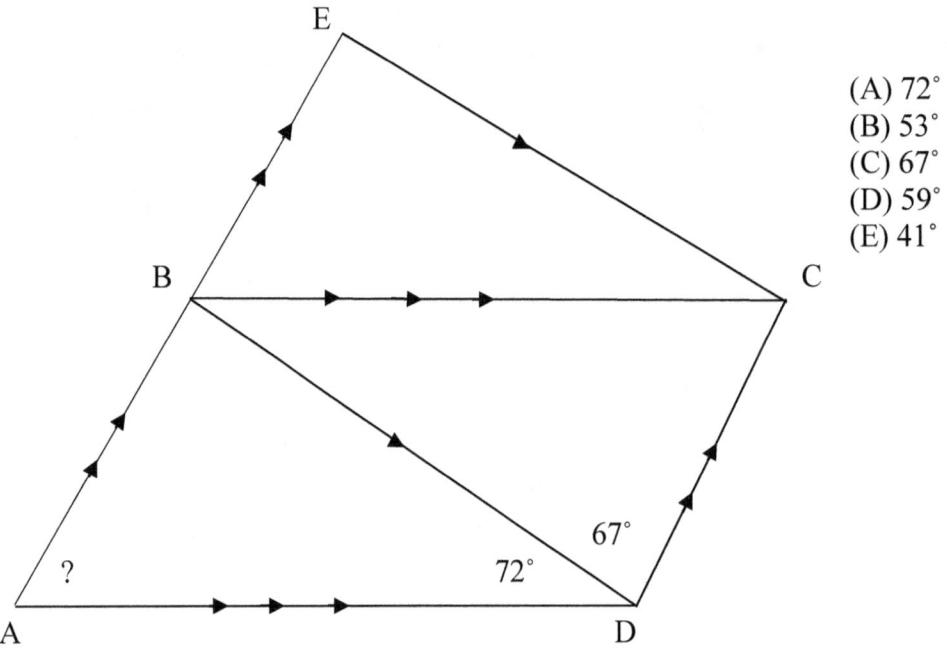

(A) 72°
(B) 53°
(C) 67°
(D) 59°
(E) 41°

$m\angle E = m\angle BDC = 67°$ because opposite angles of a parallelogram are congruent.

$m\angle ABD = m\angle BDC$ because alternate interior angles of parallel lines are congruent.

$m\angle A = 180° - 72° - 67° = 41°$

Answer: (E) 41°

12. Divide:

$$\frac{8x^3 + 2x^2 - 11x + 6}{2x + 3}$$

(A) $4x^2 + x + 3$ (B) $4x^3 + x^2 - 6$ (C) $4x^2 - 5x + 2$ (D) $4x^2 + x + 4$ (E) $4x^2 + 4x + 3$

Just divide by long division.

12. Divide:

$$\frac{8x^3 + 2x^2 - 11x + 6}{2x + 3}$$

(A) $4x^2 + x + 3$ (B) $4x^3 + x^2 - 6$ (C) $4x^2 - 5x + 2$ (D) $4x^2 + x + 4$ (E) $4x^2 + 4x + 3$

$$\begin{array}{r} 4x^2 - 5x + 2 \\ 2x+3 \overline{\smash{\big)} 8x^3 + 2x^2 - 11x + 6} \\ \underline{8x^3 + 12x^2} \\ -10x^2 - 11x + 6 \\ \underline{-10x^2 - 15x} \\ 4x + 6 \\ \underline{4x + 6} \\ 0 \end{array}$$

Answer: (C) $4x^2 - 5x + 2$

13. Find the midpoint of the line connecting (a, b) and $(a - 4, b + 6)$.

(A) $(a - 2, b + 3)$ (B) $(a + 3, b - 2)$ (C) $(a + 3, b + 3)$ (D) $(a - 2, b - 1)$
(E) $(a - 2, b + 1)$

Add the *x*-coordinates and divide by 2 to get the abscissa of the midpoint. For the ordinate of the midpoint, add the *y*-coordinates and divide by 2.

13. Find the midpoint of the line connecting (a, b) and $(a - 4, b + 6)$.

(A) $(a - 2, b + 3)$ (B) $(a + 3, b - 2)$ (C) $(a + 3, b + 3)$ (D) $(a - 2, b - 1)$
(E) $(a - 2, b + 1)$

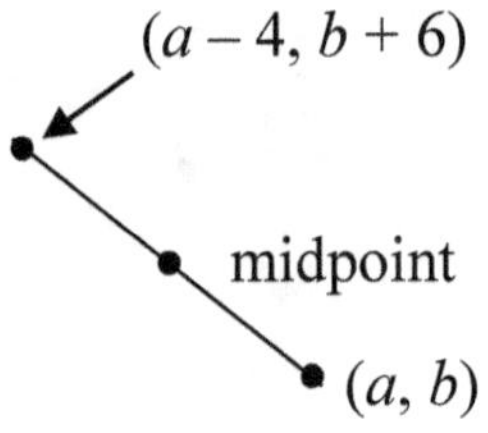

x midpoint: $\dfrac{a+(a-4)}{2} = \dfrac{2a-4}{2} = a - 2$

y midpoint: $\dfrac{b+(b+6)}{2} = \dfrac{2b+6}{2} = b + 3$

Midpoint: $(a - 2, b + 3)$

Answer: (A) $(a - 2, b + 3)$

14. ABCDEF is a regular hexagon. Diagonal AD = 8. Find the perimeter of the hexagon.

(A) 12   (B) 24   (C) 20   (D) 30   (E) 18

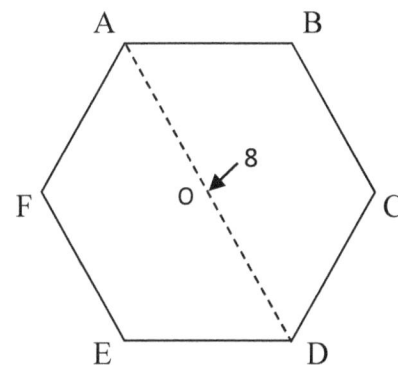

Construct a right triangle by drawing a perpendicular to side AB of the hexagon. To find a side of this new Δ, draw a unit 30°–60°–90° Δ and compare the sides. Double the side of the Δ within the hexagon to get a full side of the hexagon and then multiply by 6.

14. ABCDEF is a regular hexagon. Diagonal AD = 8. Find the perimeter of the hexagon.

(A) 12 (B) 24 (C) 20 (D) 30 (E) 18

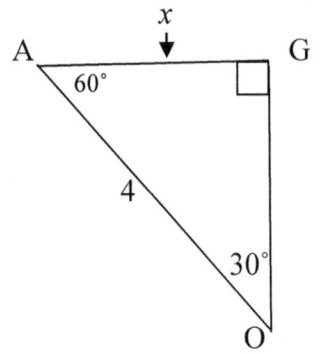

ABCDEF is a regular hexagon, so each central angle is $\frac{360}{6} = 60°$. The diagonals bisect each other, so AO = 4. Draw OG ⊥ AB, so that we have right △ AGO. Now, $m\angle AOG = 30°$ and $m\angle GAO = 60°$.

△ AGO is a 30°–60°–90° △, so we'll compare it to the unit △.

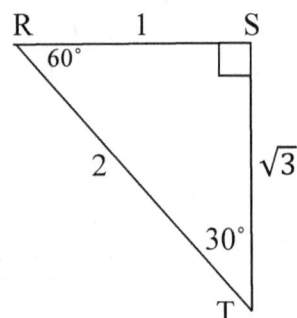

$\dfrac{x}{4} = \dfrac{1}{2}$

$2x = 4$
$x = 2$
AG = 2
AB = 4
perimeter = 6 • 4 = 24

Answer: (B) 24

15. If the second prime number after 17 is represented as $8^n + k$, where $n = 2/3$ and k is a positive integer, find k.

(A) 19 (B) 21 (C) 16 (D) 17 (E) 7

Find the second prime number after 17, set it equal to $8^n + k$ and substitute $\frac{2}{3}$ for n.

16. The annual tuition at University A is $\$r$ and the annual tuition at University B is $\$s$. If Malcolm attends University A for 3 years and then switches to University B for his last year, what is his average tuition per year over 4 years?

(A) $\dfrac{\$4 + \$s}{4}$ (B) $\dfrac{\$3r + \$1}{3}$ (C) $\dfrac{\$4r + \$s}{4}$ (D) $\dfrac{\$3r + \$s}{4}$ (E) $\dfrac{\$2r + \$s}{4}$

Multiply the number of years at each university by the annual tuition, and then add the two figures and divide by the total number of years at the two universities.

15. If the second prime number after 17 is represented as $8^n + k$, where $n = 2/3$ and k is a positive integer, find k.

(A) 19   (B) 21   (C) 16   (D) 17   (E) 7

The second next prime number after 17 is 23. Since this is represented as $8^n + k$, we get

$$8^n + k = 23$$

Since $n = 2/3$, this becomes

$$8^{2/3} + k = 23$$
$$4 + k = 23$$
$$k = 19$$

Answer: (A) 19

16. The annual tuition at University A is \$$r$ and the annual tuition at University B is \$$s$. If Malcolm attends University A for 3 years and then switches to University B for his last year, what is his average tuition per year over 4 years?

(A) $\dfrac{\$4 + \$s}{4}$ (B) $\dfrac{\$3r + \$1}{3}$ (C) $\dfrac{\$4r + \$s}{4}$ (D) $\dfrac{\$3r + \$s}{4}$ (E) $\dfrac{\$2r + \$s}{4}$

3 year's tuition at University A: \$$3r$
1 year's tuition at University B: \$$1s$

Average tuition over 4 years:

$$\dfrac{\$3r + \$s}{4}$$

Answer: (D) $\dfrac{\$3r+\$s}{4}$

17. There are 5 yellow balls, 4 white balls and 7 green balls in a container. The first ball drawn at random from the container is green and is not returned to the container. What are the odds of a second drawn ball being either yellow or white?

(A) 2/5   (B) 3/5   (C) 4/7   (D) 3/7   (E) 4/5

Determine the total number of balls remaining after a green ball is chosen. Then add the probabilities of choosing a yellow or a white ball from the total of the remaining balls.

18. Andrew takes 1/3 of the apples in a basket. Shanequa takes 1/3 of the remaining apples, and Kurt takes 1/3 of the remaining apples after that. If there are 16 apples left in the basket, how many apples were there originally?

(A) 36   (B) 54   (C) 96   (D) 68   (E) 46

Let y = the original number of apples in the basket. Keep subtracting 1/3 of the remaining apples and set the final result equal to 16.

17. There are 5 yellow balls, 4 white balls and 7 green balls in a container. The first ball drawn at random from the container is green and is not returned to the container. What are the odds of a second drawn ball being either yellow or white?

(A) 2/5   (B) 3/5   (C) 4/7   (D) 3/7   (E) 4/5

First Ball: If 1 green ball is drawn, then there are 6 green balls left.

There are still 5 yellow balls.
(+) And there are 4 white balls.
New total: 15 balls

Second Ball: So, the probability of selecting a yellow ball is $\frac{5}{15}$.

(+) the probability of selecting a white ball is $\frac{4}{15}$.

Probability of either yellow or white ball: $\frac{9}{15} = \frac{3}{5}$

Answer: (B) $\frac{3}{5}$

18. Andrew takes 1/3 of the apples in a basket. Shanequa takes 1/3 of the remaining apples, and Kurt takes 1/3 of the remaining apples after that. If there are 16 apples left in the basket, how many apples were there originally?

(A) 36   (B) 54   (C) 96   (D) 68   (E) 46

Let y = the original number of apples in the basket.

Let $\frac{1}{3}y$ = the number of apples Andrew takes, and $\frac{2}{3}y$ remains in the basket.

Let $\frac{1}{3} \cdot \frac{2}{3}y = \frac{2}{9}y$ = Shanequa's share, and $y - \frac{1}{3}y - \frac{2}{9}y = \frac{4}{9}y$ remains in the basket.

Let $\frac{1}{3} \cdot \frac{4}{9}y = \frac{4}{27}y$ = Kurt's share, and $y - \frac{1}{3}y - \frac{2}{9}y - \frac{4}{27}y = \frac{8}{27}y$ remains in the basket.

$$\frac{8}{27}y = 16$$
$$y = 54$$

Answer: (B) 54

19. If x is a number between 3 and 4, locate $(x-1)^2$.

(A) $4 < (x-1)^2 < 9$
(B) $16 < (x-1)^2 < 25$
(C) $9 \leq (x-1)^2 \leq 16$
(D) $16 \leq (x-1)^2 \leq 25$
(E) $-2 < (x-1)^2 < 2$

Let x be a number between 3 and 4. Subtract 1 and square the result.

20. Given isosceles triangle ABC, with AB ≅ BC. If m∠B = 32°, find m∠1.

(A) 32°   (B) 106°   (C) 74°   (D) 64°   (E) 148°

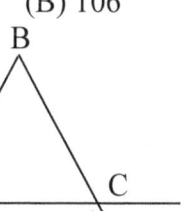

The base angles of an isosceles triangle are congruent.

19. If x is a number between 3 and 4, locate $(x-1)^2$.

(A) $4 < (x-1)^2 < 9$ (B) $16 < (x-1)^2 < 25$ (C) $9 \leq (x-1)^2 \leq 16$
(D) $16 \leq (x-1)^2 \leq 25$ (E) $-2 < (x-1)^2 < 2$

We are given that x is between 3 and 4:

$$3 < x < 4$$

Now, our goal is to build the expression $(x-1)^2$ out of this inequality. Subtracting 1 from each term of the inequality yields

$$3 - 1 < x - 1 < 4 - 1$$
$$2 < x - 1 < 3$$

Now, squaring each term of the inequality (which preserves the direction of the inequalities because all the terms are positive) yields

$$2^2 < (x-1)^2 < 3^2$$
$$4 < (x-1)^2 < 9$$

Answer: (A) $4 < (x-1)^2 < 9$

20. Given isosceles triangle ABC, with AB ≅ BC. If m∠B = 32°, find m∠1.

(A) 32°   (B) 106°   (C) 74°   (D) 64°   (E) 148°

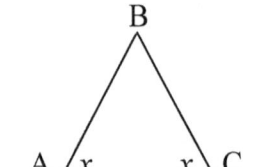

Let the equal base angles be x, as shown in the figure. Since the angle sum of a triangle is 180°, we get

$$x + x + m\angle B = 180°$$
$$x + x + 32° = 180°$$
$$2x = 148°$$
$$x = 74°$$
$$m\angle 1 + 74° = 180°$$
$$m\angle 1 = 106°$$

Answer: (B) 106°

21. If $f(x) = 3x^2 - 2c + 4$ and $f(3) = 10$, find the value of c.

(A) 8 (B) $3\frac{1}{2}$ (C) $4\frac{1}{4}$ (D) $5\frac{2}{3}$ (E) $10\frac{1}{2}$

Substitute 3 for x.

22. Select the equation that best represents the graph shown. Each line on the y-axis represents one unit.

(A) $f(x) = x + 3$ (B) $f(x) = 2 + x$ for $x \geq 3$ (C) $f(x) = 2x$ for $x \geq 0$
(D) $f(x) = 3x$ for $x \geq 0$ (E) $f(x) = 3 + x$ for $x \geq 0$

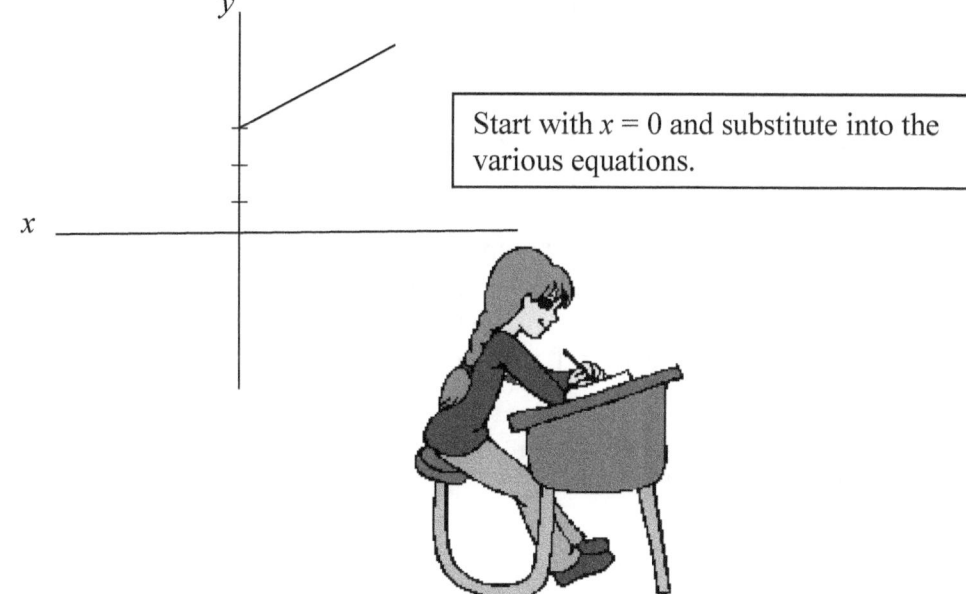

Start with $x = 0$ and substitute into the various equations.

21. If $f(x) = 3x^2 - 2c + 4$ and $f(3) = 10$, find the value of c.

(A) 8 (B) $3\frac{1}{2}$ (C) $4\frac{1}{4}$ (D) $5\frac{2}{3}$ (E) $10\frac{1}{2}$

$x = 3$:
$f(3) = 10$:

$f(x) = 3x^2 - 2c + 4$
$f(3) = 3(3)^2 - 2c + 4$
$10 = 27 - 2c + 4$
$10 = 31 - 2c$
$-2c = -21$
$c = 10\frac{1}{2}$

Answer: (E) $10\frac{1}{2}$

22. Select the equation that best represents the graph shown. Each line on the y-axis represents one unit.

(A) $f(x) = x + 3$ (B) $f(x) = 2 + x$ for $x \geq 3$ (C) $f(x) = 2x$ for $x \geq 0$
(D) $f(x) = 3x$ for $x \geq 0$ (E) $f(x) = 3 + x$ for $x \geq 0$

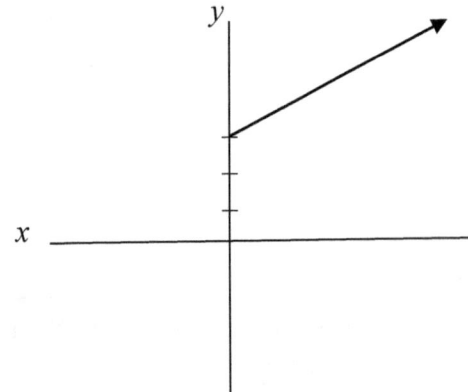

We can eliminate (A) because the graph ends at $x = 0$.
We can also eliminate (B) because the line would only begin at $x = 2 + 3 = 5$.
(C) is eliminated because at $x = 0$, $y = 0$ (but on the graph $y = 3$).
(D) is eliminated because at $x = 0$, $y = 0$ (but on the graph $y = 3$).
(E) is left standing.

Answer: (E) $f(x) = 3 + x$ for $x \geq 0$

23. Find the value of $\dfrac{3x-4}{x^3}$ when $x = 2/3$.

(A) $4\dfrac{2}{3}$ (B) $-4\dfrac{5}{6}$ (C) $-3\dfrac{4}{5}$ (D) $-6\dfrac{3}{4}$ (E) $6\dfrac{2}{3}$

Substitute 2/3 for x.

24. Two-fifths of the books printed by Kansas City Press are mysteries, one-third are biographies and one-sixth are histories. If the rest are science books and Kansas City Press publishes 60 mysteries a year, how many science books are published?

(A) 15 (B) 20 (C) 30 (D) 25 (E) 18

Find the fraction of science books published by adding all the fractions and subtracting from 1. Then, in order to discover the total number of books published, set 2/5 of the total, x, equal to 60. Finally, multiply the fraction of science books published by the total number of books published.

ACT Math Personal Tutor

23. Find the value of $\dfrac{3x-4}{x^3}$ when $x = 2/3$.

(A) $4\dfrac{2}{3}$ (B) $-4\dfrac{5}{6}$ (C) $-3\dfrac{4}{5}$ (D) $-6\dfrac{3}{4}$ (E) $6\dfrac{2}{3}$

$$\frac{3x-4}{x^3} = \frac{3\left(\frac{2}{3}\right)-4}{\left(\frac{2}{3}\right)^3} = \frac{2-4}{\frac{8}{27}} = \frac{-2}{\frac{8}{27}} = -2\cdot\frac{27}{8} = -\frac{27}{4} = -6\frac{3}{4}$$

Answer: (D) $-6\dfrac{3}{4}$

24. Two-fifths of the books printed by Kansas City Press are mysteries, one-third are biographies and one-sixth are histories. If the rest are science books and Kansas City Press publishes 60 mysteries a year, how many science books are published?

(A) 15 (B) 20 (C) 30 (D) 25 (E) 18

We'll just add up the fractions and subtract from 100% (= 1). The result is the fraction of science books.

$$\frac{2}{5} + \frac{1}{3} + \frac{1}{6} = \frac{27}{30}$$

Science books:

$$1 - \frac{27}{30} = \frac{30-27}{30} = \frac{3}{30} = \frac{1}{10}$$

Let x = the total number of books published and then set the fraction of mysteries published equal to 2/5 of the total:

$$\frac{2}{5}x = 60$$

$$x = 150$$

The total number of books: $x = 150$

To find the number of science books published, multiply 1/10 by the total, 150.

$$\frac{1}{10} \cdot 150 = 15$$

Answer: (A) 15

25. If $f(x) = x^2$ and $g(x) = 2x + 3$, find $f(g(x))$.

(A) $4x^2 + 12x + 9$ (B) $6x^2 + 9x + 4$ (C) $2x^4 + 9$ (D) $2x^4 + 3$ (E) $4x^2 + 6x + 9$

Replace x in $f(x)$ with $2x + 3$.

26. If the radius of the outer circle shown is 9 and the radius of the inner circle is 7, find the area of the shaded area. Leave your answer in terms of π.

(A) 16π (B) 32π (C) 14π (D) 24π (E) 28π

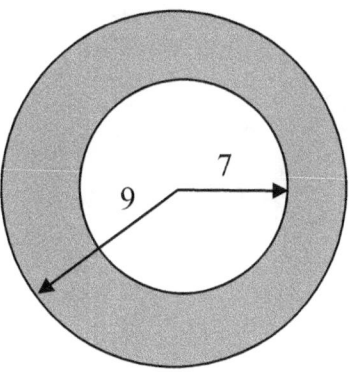

Subtract the area of the smaller circle from the area of the larger circle. Use the formula $A = \pi r^2$.

25. If $f(x) = x^2$ and $g(x) = 2x + 3$, find $f(g(x))$.

(A) $4x^2 + 12x + 9$ (B) $6x^2 + 9x + 4$ (C) $2x^4 + 9$ (D) $2x^4 + 3$ (E) $4x^2 + 6x + 9$

$$g(x) = 2x + 3$$
$$f(g(x)) = (2x + 3)^2$$
$$= 4x^2 + 12x + 9$$

Answer: (A) $4x^2 + 12x + 9$

26. If the radius of the outer circle shown is 9 and the radius of the inner circle is 7, find the area of the shaded area. Leave your answer in terms of π.

(A) 16π (B) 32π (C) 14π (D) 24π (E) 28π

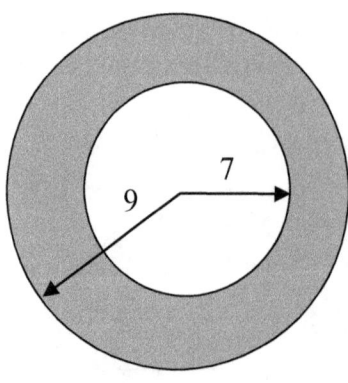

Recall that the area of a circle is πr^2:

$$A_{\text{outer circle}} - A_{\text{inner circle}} =$$

$$\pi \cdot 9^2 - \pi \cdot 7^2 =$$

$$81\pi - 49\pi =$$

$$32\pi$$

Answer: (B) 32π

27. There are 22 students in a class. On a recent exam the average grade was 83. If the lowest exam is dropped, the average goes up to 84. What was the lowest grade on the exam?

(A) 68   (B) 76   (C) 62   (D) 78   (E) 54

> Multiply the number of students in the first group by the average grade. Multiply the number of students in the second group by the average grade and then subtract the two products.

28. Calculate x in the equation shown.

$$x = \left\{\left(4 - \frac{6}{\sqrt{5}}\right)^{1/2} - \left(4 + \frac{6}{\sqrt{5}}\right)^{1/2}\right\} \cdot \left\{\left(4 - \frac{6}{\sqrt{5}}\right)^{1/2} + \left(4 + \frac{6}{\sqrt{5}}\right)^{1/2}\right\}$$

(A) $\dfrac{2}{\sqrt{5}}$ (B) $\dfrac{3\sqrt{2}}{\sqrt{3}}$ (C) $\dfrac{-2\sqrt{3}}{\sqrt{2}}$ (D) $\dfrac{-12}{\sqrt{5}}$ (E) $\dfrac{3\sqrt{5}}{2}$

> $(a - b)(a + b) = a^2 - b^2$

ACT Math Personal Tutor

27. There are 22 students in a class. On a recent exam the average grade was 83. If the lowest exam is dropped, the average goes up to 84. What was the lowest grade on the exam?

(A) 68 (B) 76 (C) 62 (D) 78 (E) 54

First, multiply 22 exams by the average grade, 83. If one exam is dropped, there are 21 exams. Multiply 21 by the average grade, 84. Then subtract the two products to find the lowest grade on the exam.

$$\begin{array}{r} \text{Total scores of 22 students} = 22 \times 83 = 1826 \\ (-)\quad \text{Total scores of 21 students} = 21 \times 84 = 1764 \\ \hline \text{Lowest grade} = \qquad\qquad\quad 62 \end{array}$$

Answer: (C) 62

28. Calculate x in the equation shown.

$$x = \left\{\left(4 - \frac{6}{\sqrt{5}}\right)^{1/2} - \left(4 + \frac{6}{\sqrt{5}}\right)^{1/2}\right\} \cdot \left\{\left(4 - \frac{6}{\sqrt{5}}\right)^{1/2} + \left(4 + \frac{6}{\sqrt{5}}\right)^{1/2}\right\}$$

(A) $\dfrac{2}{\sqrt{5}}$ (B) $\dfrac{3\sqrt{2}}{\sqrt{3}}$ (C) $\dfrac{-2\sqrt{3}}{\sqrt{2}}$ (D) $\dfrac{-12}{\sqrt{5}}$ (E) $\dfrac{3\sqrt{5}}{2}$

Using FOIL multiplication gives

$$x = \left(4 - \frac{6}{\sqrt{5}}\right) + \left(4 - \frac{6}{\sqrt{5}}\right)^{\frac{1}{2}} \cdot \left(4 + \frac{6}{\sqrt{5}}\right)^{\frac{1}{2}} - \left(4 + \frac{6}{\sqrt{5}}\right)^{\frac{1}{2}} \cdot \left(4 - \frac{6}{\sqrt{5}}\right)^{\frac{1}{2}} - \left(4 + \frac{6}{\sqrt{5}}\right)$$

$$x = \frac{-12}{\sqrt{5}}$$

Answer: (D) $\dfrac{-12}{\sqrt{5}}$

Method II: Using the Difference of Squares shortcut, $(a - b)(a + b) = a^2 - b^2$, we merely square the first term, $\left(4 - \dfrac{6}{\sqrt{5}}\right)^{1/2}$, and then subtract the square of the second term, $\left(4 + \dfrac{6}{\sqrt{5}}\right)^{1/2}$:

$$\left[\left(4 - \frac{6}{\sqrt{5}}\right)^{\frac{1}{2}}\right]^2 - \left[\left(4 + \frac{6}{\sqrt{5}}\right)^{\frac{1}{2}}\right]^2 =$$

$$\left(4 - \frac{6}{\sqrt{5}}\right) - \left(4 + \frac{6}{\sqrt{5}}\right) =$$

$$4 - \frac{6}{\sqrt{5}} - 4 - \frac{6}{\sqrt{5}} =$$

$$\frac{-12}{\sqrt{5}}$$

29. What is the maximum rectangular area 600 feet of fencing can enclose?

(A) 15,500 sq. ft.   (B) 12,000 sq. ft.   (C) 18,500 sq. ft.   (D) 22,500 sq. ft.
(E) 25,000 sq. ft.

Draw a rectangle and label the sides. Then develop a formula, $y = ax^2 + bx + c$, for the area, which is represented graphically by a parabola. If the coefficient of x^2 is negative, then the maximum height is at $x = b/2a$. Substitute that value into the width and length in order to find its maximum area.

29. What is the maximum rectangular area 600 feet of fencing can enclose?

(A) 15,500 sq. ft.   (B) 12,000 sq. ft.   (C) 18,500 sq. ft.   (D) 22,500 sq. ft.
(E) 25,000 sq. ft.

Let the length of the rectangle shown be y, and let the width be x. Since we are given 600 feet of fencing, the perimeter ($P = 2l + 2w$) is 600:

$$2l + 2w = 600$$

$$2y + 2x = 600$$

$$2y = 600 - 2x$$

$$y = 300 - x$$

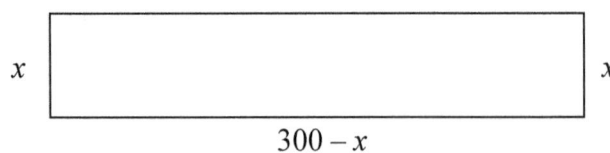

Area: (length)(width) = $(300 - x)x$
$A(x) = -x^2 + 300x$

In the equation $A(x) = -x^2 + 300x$, the coefficient of x^2 is negative, so we have a maximum height at $x = -b/2a$.

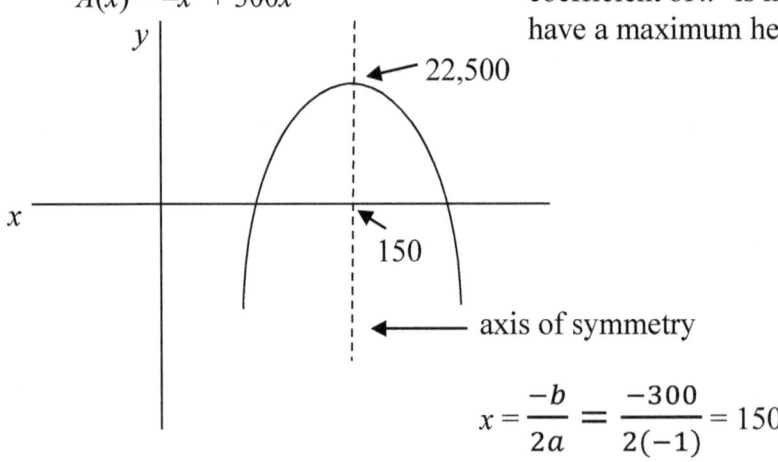

$$x = \frac{-b}{2a} = \frac{-300}{2(-1)} = 150$$

We now know that $x = 150$. Let's find $300 - x$.

width: $x = 150$
length: $300 - x = 150$
Area: $150 \cdot 150 = 22{,}500$

Answer: (D) 22,500 sq. ft.

30. The population of Tannersville has been increasing by 10% each year. If there are currently 30,000 residents, how many residents were there 2 years ago? Round off to the nearest integer.

(A) 24,793   (B) 23,732   (C) 26,873   (D) 25,985   (E) 25,980

$a_n = a_1 r^{n-1}$

30. The population of Tannersville has been increasing by 10% each year. If there are currently 30,000 residents, how many residents were there 2 years ago? Round off to the nearest integer.

(A) 24,793   (B) 23,732   (C) 26,873   (D) 25,985   (E) 25,980

"Two years ago" means that there are <u>three terms</u> in the geometric progression because we're including this year as one term.

Let a_n = the last number in a geometric progression, 30,000.
Let a_1 = the first number in the geometric progression.

Let r = the common ratio, 1.1.*
Let n = the number of terms in the geometric progression, 3.

Use the formula for the nth term of a geometric progression.

$$a_n = a_1 r^{n-1}$$

$$30{,}000 = a_1(1.1)^{3-1}$$

$$30{,}000 = a_1(1.1)^2$$

$$30{,}000 = 1.21 a_1$$

$$a_1 = 24{,}793$$

Answer: (A) 24,793

---

* After one year, the population is $30{,}000 + 10\% \cdot 30{,}000 = 30{,}000 + 3{,}000 = 33{,}000$. Now, forming the common ratio with the current population gives

$$\frac{Population\ after\ one\ year}{Current\ population} = \frac{33{,}000}{30{,}000} = 1.1$$

31. Fermat's Number, $F(n) = 2^{2^n} + 1$, produces prime numbers. If $2^{2^n} + 1 = 257$ (a prime number), find n.

(A) 2   (B) 4   (C) 3   (D) 5   (E) 6

Write each side of the equation in terms of the same base (2) and then set the exponents equal.

32. Determine the nature of the roots of the equation $y = x^2 - 6x + 9$.

(A) real, rational, unequal   (B) imaginary   (C) real, rational, equal
(D) real, irrational, equal   (E) real, irrational, unequal

Use the quadratic formula,

$$x = \frac{-b \pm \sqrt{b^2 - 4ac}}{2a}$$

ACT Math Personal Tutor

31. Fermat's Number, $F(n) = 2^{2^n} + 1$, produces prime numbers. If $2^{2^n} + 1 = 257$ (a prime number), find n.

(A) 2   (B) 4   (C) 3   (D) 5   (E) 6

$$2^{2^n} + 1 = 257$$
$$2^{2^n} = 256$$
$$2^{2^n} = 2^8$$
$$2^n = 8$$
$$2^n = 2^3$$
$$n = 3$$

Answer: (C) 3

32. Determine the nature of the roots of the equation $y = x^2 - 6x + 9$.

(A) real, rational, unequal   (B) imaginary   (C) real, rational, equal
(D) real, irrational, equal   (E) real, irrational, unequal

Using the quadratic formula yields

$$x = \frac{-b \pm \sqrt{b^2 - 4ac}}{2a} =$$

$a = 1, b = -6, c = 9$:

$$\frac{-(-6) \pm \sqrt{(-6)^2 - 4(1)(9)}}{2(1)} =$$

$$\frac{6 \pm \sqrt{36 - 36}}{2} =$$

$$\frac{6 \pm \sqrt{0}}{2} =$$

3

$x_1 = 3$ $x_2 = 3$

Answer: (C) real, rational, equal

Note: We did not actually need the whole quadratic formula to solve this problem, just the discriminant: $b^2 - 4ac$:

$$b^2 - 4ac = \begin{cases} \text{zero,} & \text{real, equal} \\ \text{positive,} & \text{real, unequal} \\ \text{negative,} & \text{imaginary, unequal} \end{cases}$$

33. Julie can paint a house by herself is 12 days. Melissa can do the entire job alone in 6 days. Julie works on her own for 3 days and is then joined by Melissa. The two women then work together and complete the job. How many days does it take the two women to finish painting the house together.

(A) 3　　　(B) 5　　　(C) 6　　　(D) 4　　　(E) 7

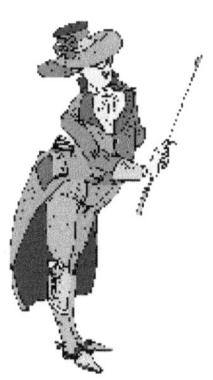

Determine what fraction of the job each woman can do in one day. Julie works alone for three days. Then the women work together for x days. Add the three fractions and set the total equal to one entire job.

34. Select a rational number between $\sqrt{5}$ and $\sqrt{6}$.

(A) $2\frac{1}{2}$　　(B) $2\frac{1}{5}$　　(C) $2\frac{3}{5}$　　(D) $2\frac{4}{5}$　　(E) $2\frac{2}{5}$

Square the given radicals. Then change the answer-choices to decimal numbers and square them.

33. Julie can paint a house by herself is 12 days. Melissa can do the entire job alone in 6 days. Julie works on her own for 3 days and is then joined by Melissa. The two women then work together and complete the job. How many days does it take the two women to finish painting the house together.

(A) 3 (B) 5 (C) 6 (D) 4 (E) 7

Julie can paint the house on her own in 12 days, so in 1 day, she does 1/12 of the job. Julie works on her own for 3 days, so she does 3/12 of the job.

In 6 days alone, Melissa can complete the job. So, in 1 day, she does 1/6 of the job. The two women then work together for x days to complete the job.

Julie works on her own for 3 days → $\frac{3}{12}$

Julie works with Melissa for x days → $\frac{x}{12}$

Melissa works with Julie for x days → $\frac{x}{6}$

$$\frac{3}{12} + \frac{x}{12} + \frac{x}{6} = 1$$

One complete job

Multiply by 12 to clear the fractions:

$$3 + x + 2x = 12$$
$$3x = 9$$
$$x = 3$$

The two women work together for 3 days.

Answer: (A) 3

34. Select a rational number between $\sqrt{5}$ and $\sqrt{6}$.

(A) $2\frac{1}{2}$ (B) $2\frac{1}{5}$ (C) $2\frac{3}{5}$ (D) $2\frac{4}{5}$ (E) $2\frac{2}{5}$

Squaring $\sqrt{5}$ and $\sqrt{6}$ gives 5 and 6. So, let's square the answer-choices to see which one is between 5 and 6. Look at Choice (E). First, convert it to a decimal:

$$2\frac{2}{5} = 2.4$$

Now, square it:

$$2.4^2 = 5.76$$

Answer: (E) $2\frac{2}{5}$

35. Find the equation of the horizontal line that passes through the maximum point of the parabola represented by the equation $-2x^2 + 4x = y + 3$.

(A) $y = -2$ (B) $y = 0$ (C) $y = -3$ (D) $y = -1$ (E) $y = -2$

Rearrange the equation in the form $y = ax^2 + bx + c$. The coefficient of x^2 in the rearranged equation is negative, so that we have a maximum height at $x = -b/2a$. Substitute the value for x in the rearranged equation in order to find the value of y.

36. If $f(x) = 2x^2 - 5x - 12$, for which positive value of x is $f(x) = 0$?

(A) 5   (B) 3   (C) 6   (D) 4   (E) 2

Set the function $f(x)$ equal to zero and then solve the resulting equation.

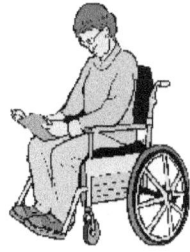

35. Find the equation of the horizontal line that passes through the maximum point of the parabola represented by the equation $-2x^2 + 4x = y + 3$.

(A) $y = -2$ (B) $y = 0$ (C) $y = -3$ (D) $y = -1$ (E) $y = -2$

$$-2x^2 + 4x = y + 3$$

$$y = -2x^2 + 4x - 3$$

The x-coordinate of the maximum point (vertex) on the parabola is given by

$$x = {-b}/{2a} = {-4}/{-4} = 1$$

Plugging $x = 1$ into the equation of the parabola gives the maximum height:

$$y = -2(1)^2 + 4(1) - 3$$

$$y = -1$$

Answer: (D) $y = -1$

36. If $f(x) = 2x^2 - 5x - 12$, for which positive value of x is $f(x) = 0$?

(A) 5   (B) 3   (C) 6   (D) 4   (E) 2

$$f(x) = 2x^2 - 5x - 12$$

$$0 = (2x + 3)(x - 4)$$

$$2x + 3 = 0 \text{ or } x - 4 = 0$$

$$x = -\frac{3}{2} \text{ or } x = 4$$
$$\text{reject}$$

Answer: (D) 4

37. Triangle ABC is inscribed in a circle. $m\angle A = 60°$ and $m\angle B = 50°$. Which of the following statements is true?

(A) $m \overset{\frown}{BC} > m \overset{\frown}{AB}$ (B) $m \overset{\frown}{AB} < m \overset{\frown}{AC}$ (C) $m \overset{\frown}{BC} < m \overset{\frown}{AC}$

(D) $m \overset{\frown}{AC} > m \overset{\frown}{BC}$ (E) $m \overset{\frown}{AB} > m \overset{\frown}{BC}$

Draw a circle and an inscribed triangle. Label the angles and determine the arcs.

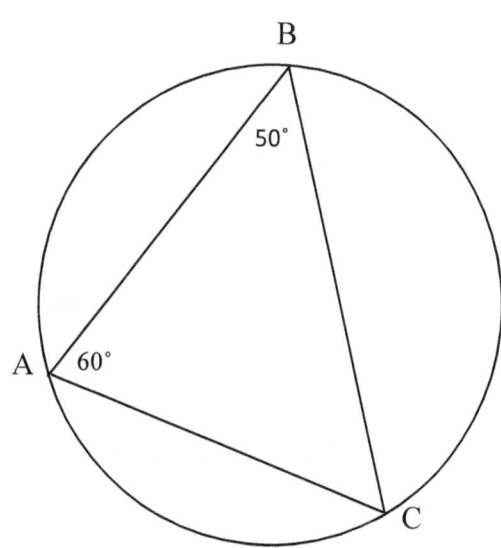

37. Triangle ABC is inscribed in a circle. $m\angle A = 60°$ and $m\angle B = 50°$. Which of the following statements is true?

(A) $m\widehat{BC} > m\widehat{AB}$ (B) $m\widehat{AB} < m\widehat{AC}$ (C) $m\widehat{BC} < m\widehat{AC}$

(D) $m\widehat{AC} > m\widehat{BC}$ (E) $m\widehat{AB} > m\widehat{BC}$

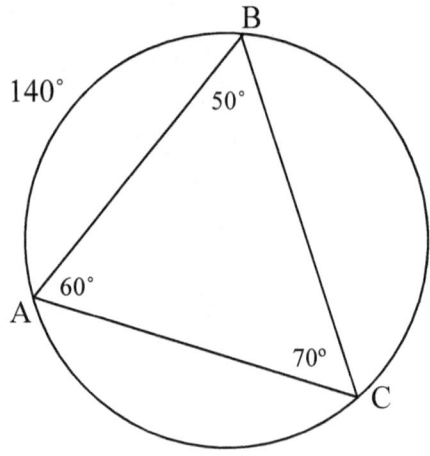

Since there are 180° in a triangle, $m\angle C + 50° + 60° = 180°$. Solving for $m\angle C$ yields

$$m\angle C = 180° - 60° - 50° = 70°$$

An inscribed angle is equal to 1/2 the measure of its intercepted arc.

$$m\widehat{AB} = 140°, m\widehat{AC} = 100°, m\widehat{BC} = 120°$$

(A) $m\widehat{BC} > m\widehat{AB}$ False

(B) $m\widehat{AB} < m\widehat{AC}$ False

(C) $m\widehat{BC} < m\widehat{AC}$ False

(D) $m\widehat{AC} > m\widehat{BC}$ False

(E) $m\widehat{AB} > m\widehat{BC}$ True

Answer: (E) $m\widehat{AB} > m\widehat{BC}$

38. Find $\tan(\arcsin \frac{n}{m})$.

(A) $\frac{\sqrt{n}}{m^2}$ (B) $\frac{n^2}{\sqrt{m}}$ (C) $\frac{n}{\sqrt{m^2 - n^2}}$ (D) $\frac{\sqrt{m^2 - n}}{m}$ (E) $\frac{mn}{\sqrt{n - m}}$

Find the tangent of an angle whose sine is *n/m*.

38. Find $\tan(\arcsin \frac{n}{m})$.

(A) $\dfrac{\sqrt{n}}{m^2}$ (B) $\dfrac{n^2}{\sqrt{m}}$ (C) $\dfrac{n}{\sqrt{m^2-n^2}}$ (D) $\dfrac{\sqrt{m^2-n}}{m}$ (E) $\dfrac{mn}{\sqrt{n-m}}$

All inverse trig functions are angles. To emphasize this, let's set $\arcsin \frac{n}{m}$ equal to θ, a variable commonly used to represent angles:

$$\theta = \arcsin \frac{n}{m}$$

By definition, this means θ is the angle whose sine is equal to $\frac{n}{m}$:

$$\sin \theta = \frac{n}{m}$$

Since $\theta = \arcsin \frac{n}{m}$, the expression we are trying to calculate, $\tan(\arcsin \frac{n}{m})$, becomes

$$\tan \theta$$

Now, we have calculated the value of $\sin \theta$, which is by definition the opposite side of an angle of a triangle divided by the hypotenuse of the triangle. And we want to calculate the value of $\tan \theta$, which is by definition the opposite side of an angle of a triangle divided by the adjacent side of the triangle. So, we need to calculate the adjacent side of the triangle:

From the drawing, use the Pythagorean Theorem to find the adjacent side x.

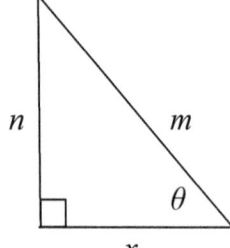

$$x^2 + n^2 = m^2$$
$$x^2 = m^2 - n^2$$
$$x = \sqrt{m^2 - n^2}$$

$$\tan \theta = \frac{opposite\ side}{adjacent\ side} = \frac{n}{\sqrt{m^2-n^2}}$$

Answer: (C) $\dfrac{n}{\sqrt{m^2-n^2}}$

39. Triangle ABC is isosceles with AC = BC. Angle ACB measures 40° and ACF is a straight line. DE || AB. Find the measure of angle FCE.

(A) 40°   (B) 70°   (C) 110°   (D) 80°   (E) 120°

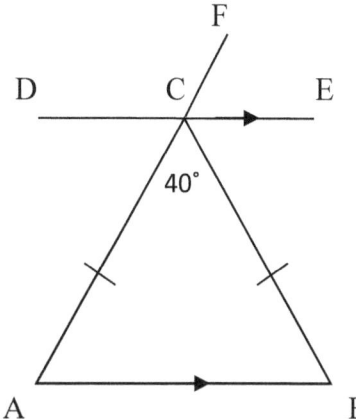

Use the properties of isosceles triangles as well as the properties of parallel lines.

39. Triangle ABC is isosceles with AC = BC. Angle ACB measures 40° and ACF is a straight line. DE ∥ AB. Find the measure of angle FCE.

(A) 40°   (B) 70°   (C) 110°   (D) 80°   (E) 120°

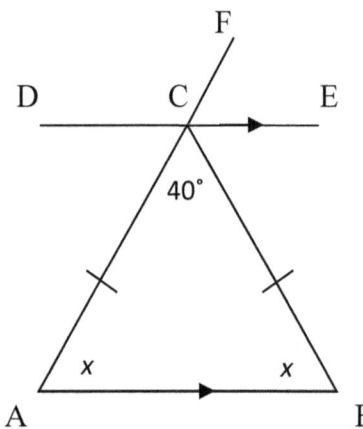

Since triangle ABC is isosceles, the base angles, A and B, are equal. Let the equal base angles be represented by x, as shown in the figure. Since the angle sum of a triangle is 180°, we get

$$x + x + 40° = 180°$$
$$2x + 40° = 180°$$
$$2x = 140°$$
$$x = 70°$$

Now, $m\angle B = m\angle BCE = 70°$ because alternate interior angles of parallel lines are congruent.

ACF is a straight line, so

$$40° + 70° + \angle FCE = 180°$$
$$110° + \angle FCE = 180°$$
$$\angle FCE = 70°$$

Note: This can be obtained more directly by noting that angles A and FCE are corresponding and therefore congruent.

Answer: (B) 70°

40. There are 10 candies in a basket. Only one of them is chocolate. Maria is permitted to roll a die. There are six sides to the die. Depending upon the number of dots the die shows, Maria is allowed that number of chances to select the chocolate. For example, if only one dot shows, Maria gets only one chance to select the chocolate. If two dots show, she gets two chances, and so forth. What are Maria's chances of selecting the chocolate?

(A) 28%   (B) 35%   (C) 42%   (D) 49%   (E) 25%

Maria's chance of selecting a chocolate is the average of all the probabilities.

40. There are 10 candies in a basket. Only one of them is chocolate. Maria is permitted to roll a die. There are six sides to the die. Depending upon the number of dots the die shows, Maria is allowed that number of chances to select the chocolate. For example, if only one dot shows, Maria gets only one chance to select the chocolate. If two dots show, she gets two chances, and so forth. What are Maria's chances of selecting the chocolate?

(A) 28%   (B) 35%   (C) 42%   (D) 49%   (E) 25%

If a single dot shows, she has 1 chance in 10 of selecting the chocolate: 1/10.

If a two dots show, she has 2 chances in 10 of selecting the chocolate: 2/10.

If a three dots show, she has 3 chances in 10 of selecting the chocolate: 3/10.

Etc.

Maria's chance of selecting a chocolate is the average of all the probabilities:

$$\frac{\frac{1}{10}+\frac{2}{10}+\frac{3}{10}+\frac{4}{10}+\frac{5}{10}+\frac{6}{10}}{6} =$$

$$\frac{1+2+3+4+5+6}{60} =$$

$$\frac{21}{60} = .35$$

Answer: (B) 35%

41. If *a* is an even negative integer and *n* is an odd negative integer and $a > n$, select the best answer to describe $(a - n)^3$.

(A) even negative integer (B) odd negative integer (C) even positive integer
(D) odd positive integer (E) even positive integer less than 25

The easiest way to solve this problem is to assign different sets of even and odd negative integers to *a* and *n* and then substitute them into $(a - n)^3$.

41. If a is an even negative integer and n is an odd negative integer and $a > n$, select the best answer to describe $(a - n)^3$.

(A) even negative integer (B) odd negative integer (C) even positive integer
(D) odd positive integer (E) even positive integer less than 25

The easiest way to solve this problem is to just select two even negative values for a, two odd negative values for n such that $a > n$, plug them into $(a - n)^3$, and then observe the result.

Case 1) Let $a = -6$, $n = -7$

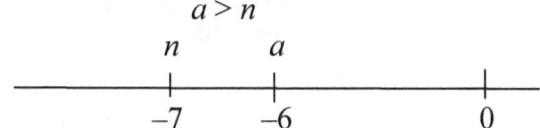

$(a - n)^3$
$(-6 - [-7])^3$
$(-6 + 7)^3$
1^3
1

Here, we have an odd positive integer, 1.

Case 2) Let $a = -2$, $n = -11$

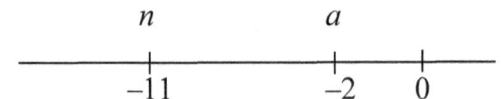

$(a - n)^3$
$(-2 - [-11])^3$
$(-2 + 11)^3$
9^3
729

In both cases the result is an odd positive integer.

Answer: (D) odd positive integer

42. If $a = 6$ and $b = 3$ in the triangle shown, what is a possible value for side c?

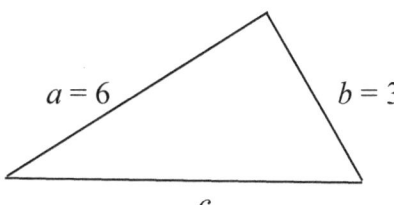

(A) 4   (B) 11   (C) 9   (D) 10   (E) 12

The sum of any two sides of a triangle is greater than the third side.

42. If $a = 6$ and $b = 3$ in the triangle shown, what is a possible value for side c?

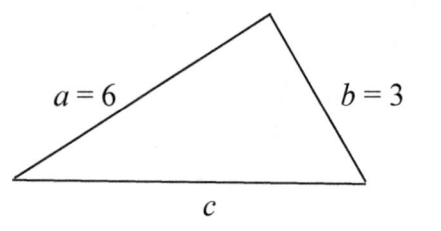

(A) 4 (B) 11 (C) 9 (D) 10 (E) 12

The sum of any two sides of a triangle must be greater than the third side. The sum of sides a and b is $6 + 3 = 9$, and the only answer-choice less than 9 is 4. Hence, the answer is (A).

The following diagrams illustrate the five answer-choices:

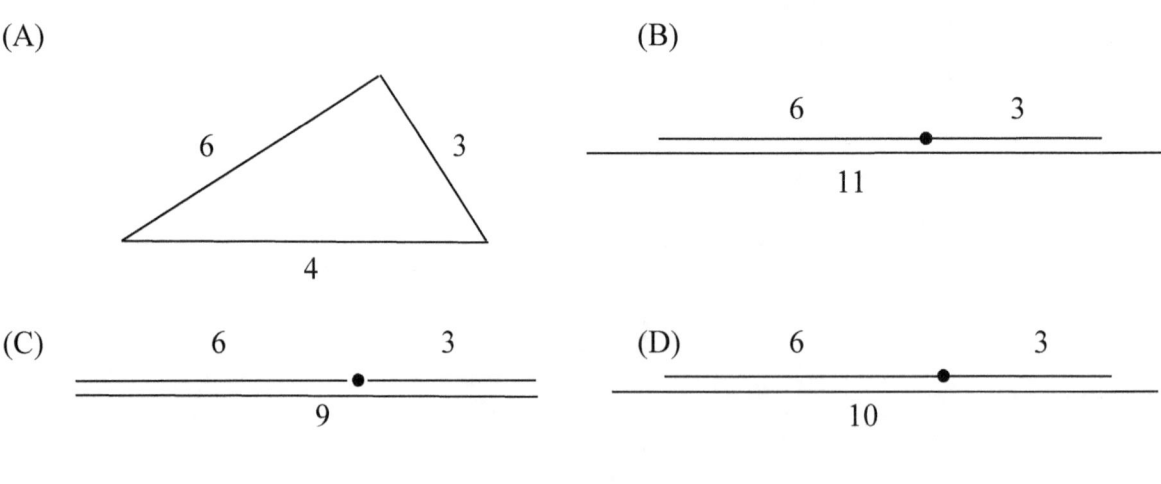

Answer: (A) 4

43. Which of the following equations are functions?

(1) $y = x + 2$ (2) $x^2 + y^2 = 25$ (3) $y = 4$ (4) $x = 3$ (5) $x = y$

(A) (1) and (3)   (B) (2) and (4)   (C) (3) and (5)   (D) (3) and (4)
(E) (1), (3) and (5)

A function is a relation describing an ordered pair in which the first element, x, is mapped into one and only one second element, y, of the ordered pair. Check to see which equations fit this definition.

vertical line test

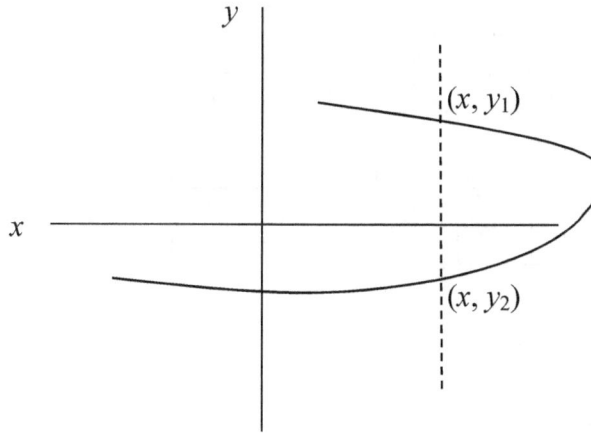

This is not a function because x is mapped into two y's.

43. Which of the following equations are functions?

(1) $y = x + 2$ (2) $x^2 + y^2 = 25$ (3) $y = 4$ (4) $x = 3$ (5) $x = y$

(A) (1) and (3)   (B) (2) and (4)   (C) (3) and (5)   (D) (3) and (4)
(E) (1), (3) and (5)

(1)

$y = x + 2$ Function	
x	y
0	2
1	3
2	4
−1	1
−2	0

$x^2 + y^2 = 25$ is not a function because x is mapped into 2 different y's.

(2)

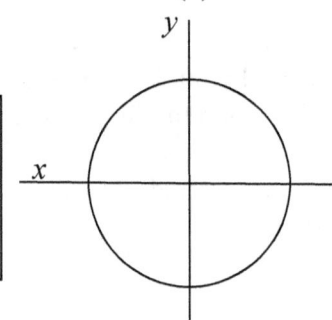

(3)

$y = 4$ Function	
x	y
0	4
1	4
2	4
−1	4
−2	4

(4)

$x = 3$ Not a Function	
x	y
3	0
3	1
3	2
3	3
3	4

(5)

$x = y$ Function	
x	y
0	0
1	1
2	2
−1	−1
−2	−2

Answer: (E) (1), (3) and (5)

44. Universal set, U, is the set of integers between and including 1 and 9. The set A is the set of even integers greater than 1 but less than 6 while set B is the set of odd integers greater than 3 and less than 10.

Find (A ∪ B) ∩ (~B).

(A) {2, 5, 8}   (B) {2, 4, 9}   (C) {2, 4}   (D) {3, 5, 7}   (E) {2, 3, 4, 5}

In the case of union (∪) of two sets, include all elements of both sets. In the case of intersection (∩) of two sets, include only elements common to both sets. The negation (~) of a set includes only elements external to that set but part of the Universal Set.

45. What is the sum of the first 25 multiples of 7?

(A) 1,980   (B) 2,275   (C) 3,850   (D) 1,725   (E) 2,275

List the first multiple of 7. Then list the 25th multiple of 7 and use the formula for the sum of an arithmetic sequence:

$$S_n = \frac{n}{2}(a_1 + a_n)$$

ACT Math Personal Tutor

44. Universal set, U, is the set of integers between and including 1 and 9. The set A is the set of even integers greater than 1 but less than 6 while set B is the set of odd integers greater than 3 and less than 10.

Find (A ∪ B) ∩ (~B).

(A) {2, 5, 8}   (B) {2, 4, 9}   (C) {2, 4}   (D) {3, 5, 7}   (E) {2, 3, 4, 5}

$$U = \{1, 2, 3, 4, 5, 6, 7, 8, 9\}$$
$$A = \{2, 4\}$$
$$B = \{5, 7, 9\}$$

$$A \cup B = \{2, 4, 5, 7, 9\}$$
$$\sim B = \{1, 2, 3, 4, 6, 8\}$$

$$(A \cup B) \cap (\sim B)$$
$$\{2, 4, 5, 7, 9\} \cap \{1, 2, 3, 4, 6, 8\}$$
$$\{2, 4\}$$

Answer: (C) {2, 4}

45. What is the sum of the first 25 multiples of 7?

(A) 1,980   (B) 2,275   (C) 3,850   (D) 1,725   (E) 2,275

$a_1 = 1 \times 7 = 7$
$a_{25} = 25 \times 7 = 175$

$$S_n = \frac{n}{2}(a_1 + a_n)$$

$n = 25$:

$$S_{25} = \frac{25}{2}(7 + 175)$$

$$S_{25} = 2275$$

Answer: (E) 2,275

46. If a parabola described by $f(x) = x^2$ is represented by the graph shown, describe the graph represented by the inverse.

(A) opens at the bottom (B) opens to the right (C) opens to the left
(D) opens at the top (E) is centered along the line $y = x$

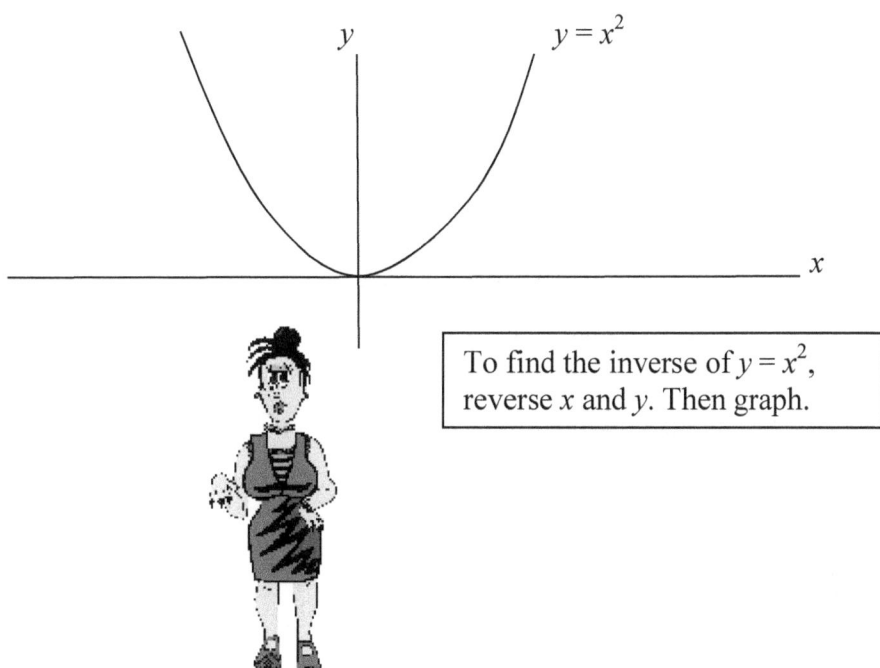

To find the inverse of $y = x^2$, reverse x and y. Then graph.

47. If $f(x, y) = 3x - 2y + 2$, find $f(f(4, -1), 3)$.

(A) 18 (B) 32 (C) 26 (D) 44 (E) 52

First, find $f(4, -1)$. Then substitute that answer into $f(x, 3)$.

46. If a parabola described by $f(x) = x^2$ is represented by the graph shown, describe the graph represented by the inverse.

(A) opens at the bottom (B) opens to the right (C) opens to the left
(D) opens at the top (E) is centered along the line $y = x$

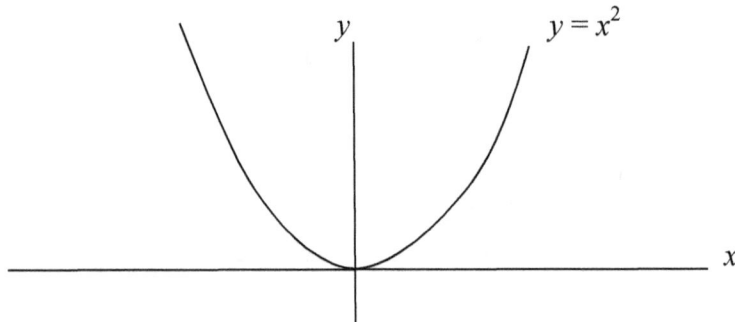

To form the inverse, reverse x and y: $\begin{aligned} y &= x^2 \\ x &= y^2 \end{aligned}$
Solve for y: $y = \pm\sqrt{x}$

It's easier to use your graphing calculator to draw the graph.

However, you can also develop a table.

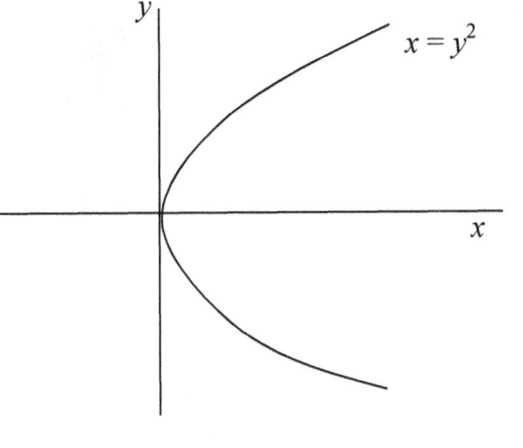

$y = \pm\sqrt{x}$

x	y
0	0
1	± 1
2	$\pm 1.4^+$
3	$\pm 1.7^+$

Answer: (B) opens to the right

47. If $f(x, y) = 3x - 2y + 2$, find $f(f(4, -1), 3)$.

(A) 18 (B) 32 (C) 26 (D) 44 (E) 52

$$f(x, y) = 3x - 2y + 2$$
$$f(4, -1) = 3(4) - 2(-1) + 2 = 16$$
$$f(16, 3) = 3(16) - 2(3) + 2 = 44$$

Answer: (D) 44

48. If $f(x) = x + 3$ and $g(x) = x^2 + 4x + 3$, for which values of x is $f(x) = g(x)$?

(A) $x = 1, 2$ (B) $x = 2, 3$ (C) $x = 1, -2$ (D) $x = -1, 2$ (E) $x = 0, -3$

Set $f(x) = g(x)$ to determine the value of x.

49. If $\csc \theta = 2$, find $\tan^2 \theta$.

(A) 1/2 (B) 1/3 (C) $\dfrac{1}{\sqrt{2}}$ (D) $\dfrac{2}{\sqrt{3}}$ (E) 4

$$\csc \theta = \frac{1}{\sin \theta}$$

48. If $f(x) = x + 3$ and $g(x) = x^2 + 4x + 3$, for which values of x is $f(x) = g(x)$?

(A) $x = 1, 2$ (B) $x = 2, 3$ (C) $x = 1, -2$ (D) $x = -1, 2$ (E) $x = 0, -3$

$$f(x) = g(x)$$
$$x + 3 = x^2 + 4x + 3$$
$$x^2 + 3x = 0$$
$$x(x + 3) = 0$$
$$x = 0 \text{ or } x = -3$$

Answer: (E) $x = 0, -3$

49. If $\csc \theta = 2$, find $\tan^2 \theta$.

(A) 1/2 (B) 1/3 (C) $\dfrac{1}{\sqrt{2}}$ (D) $\dfrac{2}{\sqrt{3}}$ (E) 4

$$\csc \theta = \frac{1}{\sin \theta} = 2$$
$$\sin \theta = \frac{1}{2}$$

Now, we have calculated the value of sin θ, which is by definition the opposite side of an angle of a triangle divided by the hypotenuse of the triangle. And we want to calculate the value of tan θ, which is by definition the opposite side of an angle of a triangle divided by the adjacent side of the triangle. So, we need to calculate the adjacent side of the triangle:

From the drawing, use the Pythagorean Theorem to find the adjacent side x.

$$x^2 + 1^2 = 2^2$$
$$x^2 = 4 - 1 = 3$$
$$x = \sqrt{3}$$

$$\tan^2 \theta = (\tan \theta)^2 = \left(\frac{\text{opposite side}}{\text{adjacent side}}\right)^2 = \left(\frac{1}{\sqrt{3}}\right)^2 = \frac{1}{3}$$

Answer: (B) 1/3

50. Determine the quadratic equation the sum of whose roots is 7 and the product of those roots is 12.

(A) $x^2 - 3x + 10 = 0$ (B) $7 = 10x - x^2$ (C) $x^2 - 7x + 12 = 0$ (D) $x^2 + 3x - 7 = 0$
(E) $10x - x^2 = 2x$

Develop two equations. In the first equation, add the two roots and set their sum equal to 7. In the second equation, set the product of the two roots equal to 12.

51. If $x^2 + 2x + 3$ represents the first odd positive integer, represent the 4th even integer greater than the first odd positive integer.

(A) $x^2 + 2x + 10$ (B) $x^2 + 2x + 5$ (C) $x^2 + 2x + 12$ (D) $x^2 + 2x + 9$
(E) $x^2 + 3x + 1$

If $x^2 + 2x + 3$ represents the first odd positive integer, add 1 to represent the first even integer and go on from there.

50. Determine the quadratic equation the sum of whose roots is 7 and the product of those roots is 12.

(A) $x^2 - 3x + 10 = 0$ (B) $7 = 10x - x^2$ (C) $x^2 - 7x + 12 = 0$ (D) $x^2 + 3x - 7 = 0$
(E) $10x - x^2 = 2x$

Let the roots of the equation be x and y. Since "the sum of whose roots is 7," we get

$$(1) \quad x + y = 7$$

Since "the product of those roots is 12," we get

$$(2) \quad x \cdot y = 12$$

Solving equation (2) for y yields $y = 12/x$. Now, substitute this expression for y in equation (1):

$$x + \frac{12}{x} = 7$$

Multiply by x to clear the fraction:

$$x^2 + 12 = 7x$$

$$x^2 - 7x + 12 = 0$$

Answer: (C) $x^2 - 7x + 12 = 0$

51. If $x^2 + 2x + 3$ represents the first odd positive integer, represent the 4th even integer greater than the first odd positive integer.

(A) $x^2 + 2x + 10$ (B) $x^2 + 2x + 5$ (C) $x^2 + 2x + 12$ (D) $x^2 + 2x + 9$
(E) $x^2 + 3x + 1$

The first odd positive integer: $x^2 + 2x + 3$
The first even integer greater than the first odd integer: $(x^2 + 2x + 3) + 1 = x^2 + 2x + 4$
The second even integer greater than first odd integer: $(x^2 + 2x + 4) + 2 = x^2 + 2x + 6$
The third even integer greater than first odd integer: $(x^2 + 2x + 6) + 2 = x^2 + 2x + 8$
The fourth even integer greater than first odd integer: $(x^2 + 2x + 8) + 2 = x^2 + 2x + 10$

Answer: (A) $x^2 + 2x + 10$

52. Multiply the two matrices shown.

$$\begin{bmatrix} 2 & 1 & 5 \\ 3 & 4 & 6 \end{bmatrix} \text{ and } \begin{bmatrix} 2 & 4 \\ 5 & 1 \\ 3 & 3 \end{bmatrix}$$

(A) $\begin{bmatrix} 23 & 15 & 51 \\ 34 & 42 & 64 \end{bmatrix}$

(B) $\begin{bmatrix} 45 & 36 \\ 57 & 19 \end{bmatrix}$

(C) $\begin{bmatrix} 13 & 15 & 51 \\ 34 & 42 & 64 \end{bmatrix}$

(D) $\begin{bmatrix} 24 & 24 \\ 44 & 34 \end{bmatrix}$

(E) $\begin{bmatrix} 23 & 15 \\ 34 & 42 \end{bmatrix}$

The rows in the first matrix must match the columns in the second matrix. Then multiply the columns by the rows.

52. Multiply the two matrices shown.

$$\begin{bmatrix} 2 & 1 & 5 \\ 3 & 4 & 6 \end{bmatrix} \text{ and } \begin{bmatrix} 2 & 4 \\ 5 & 1 \\ 3 & 3 \end{bmatrix}$$

(A) $\begin{bmatrix} 23 & 15 & 51 \\ 34 & 42 & 64 \end{bmatrix}$

(B) $\begin{bmatrix} 45 & 36 \\ 57 & 19 \end{bmatrix}$

(C) $\begin{bmatrix} 13 & 15 & 51 \\ 34 & 42 & 64 \end{bmatrix}$

(D) $\begin{bmatrix} 24 & 24 \\ 44 & 34 \end{bmatrix}$

(E) $\begin{bmatrix} 23 & 15 \\ 34 & 42 \end{bmatrix}$

$$\begin{bmatrix} 2 & 1 & 5 \\ 3 & 4 & 6 \end{bmatrix} \times \begin{bmatrix} 2 & 4 \\ 5 & 1 \\ 3 & 3 \end{bmatrix} = \begin{bmatrix} 2\cdot 2 + 1\cdot 5 + 5\cdot 3 & 2\cdot 4 + 1\cdot 1 + 5\cdot 3 \\ 3\cdot 2 + 4\cdot 5 + 6\cdot 3 & 3\cdot 4 + 4\cdot 1 + 6\cdot 3 \end{bmatrix} = \begin{bmatrix} 24 & 24 \\ 44 & 34 \end{bmatrix}$$

Answer: (D) $\begin{bmatrix} 24 & 24 \\ 44 & 34 \end{bmatrix}$

53. Change 240° to radians, correct to the nearest hundredth.

(A) 1.5 π radians   (B) 1.33 π radians   (C) 2.1 π radians
(D) .88 π radians   (E) 1.78 π radians

One radian is the measure of a central angle of a circle subtended by an arc whose length is equal to the radius of the circle. 2π radians = 360°, so set up a proportion.

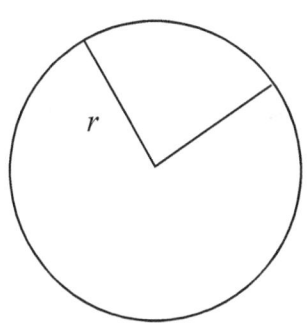

54. We have five flags of different colors. In how many different ways can we arrange the flags in a circle?

(A) 12   (B) 16   (C) 18   (D) 32   (E) 24

We can arrange n different objects in $(n-1)!$ distinct ways in a circle.

53. Change 240° to radians, correct to the nearest hundredth.

(A) 1.5 π radians (B) 1.33 π radians (C) 2.1 π radians
(D) .88 π radians (E) 1.78 π radians

$$2\pi \text{ radians} = 360°$$

Dividing both sides of this equivalence by 360° yields

$$\frac{2\pi \text{ radians}}{360°} = 1$$

This is the conversion factor for changing degrees to radians. Forming a proportion yields

$$\frac{x \text{ radians}}{240°} = \frac{2\pi \text{ radians}}{360°}$$

$$x = \frac{2\pi \cdot 240}{360°} = \frac{480\pi}{360} = \frac{4}{3}\pi \approx 1.33\pi$$

Answer: (B) 1.33 π radians

54. We have five flags of different colors. In how many different ways can we arrange the flags in a circle?

(A) 12   (B) 16   (C) 18   (D) 32   (E) 24

$$(n-1)! = (5-1)! = 4! = 4 \times 3 \times 2 \times 1 = 24$$

Answer: (E) 24

55. Using the chart shown, determine the percentage increase in the price of a pound of cherries from 2004 until 2014? Round off to the nearest percent.

(A) 200%   (B) 225%   (C) 184%   (D) 343%   (E) 156%

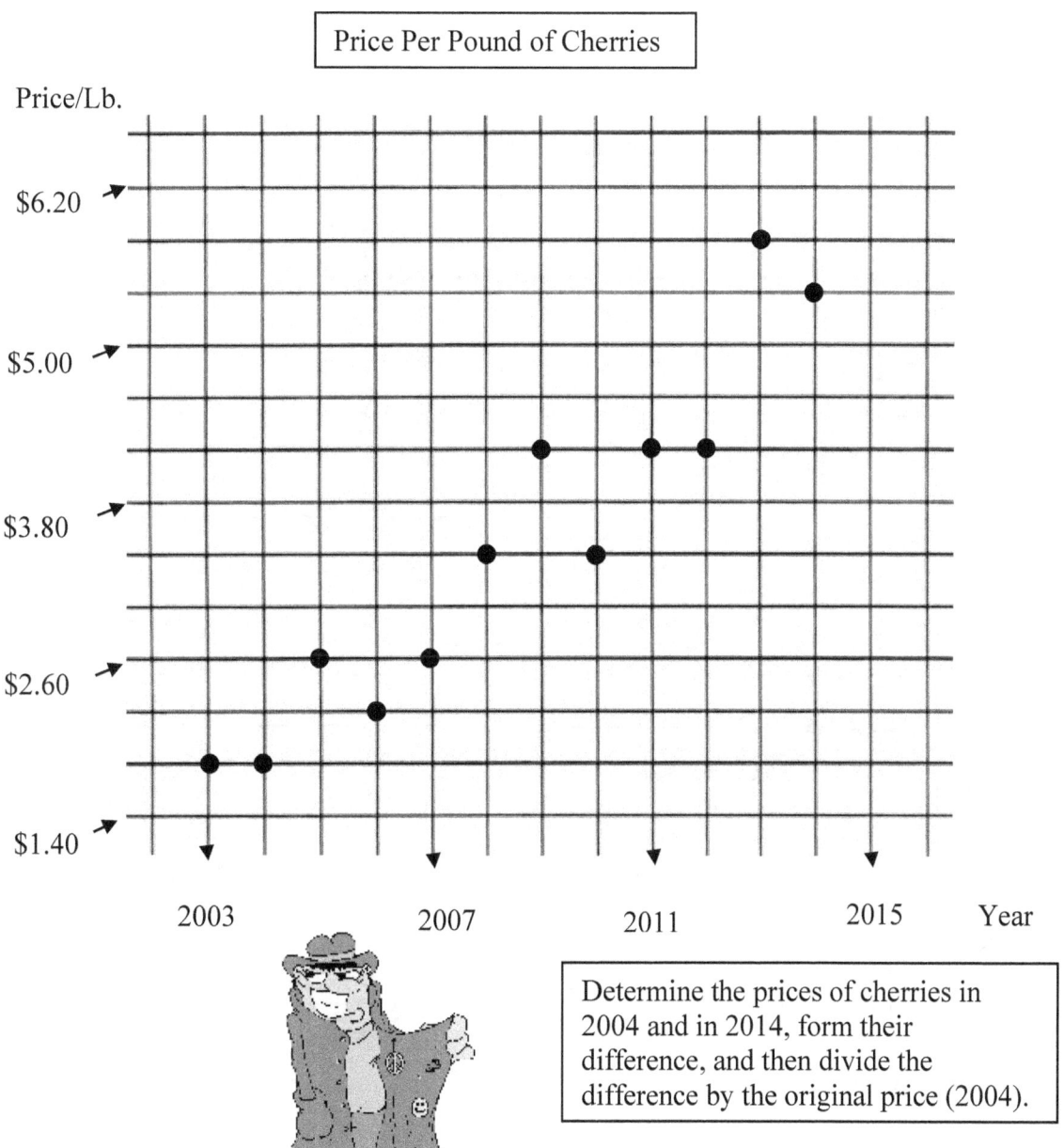

Determine the prices of cherries in 2004 and in 2014, form their difference, and then divide the difference by the original price (2004).

55. Using the chart shown, determine the percentage increase in the price of a pound of cherries from 2004 until 2014? Round off to the nearest percent.

(A) 200%   (B) 225%   (C) 184%   (D) 343%   (E) 156%

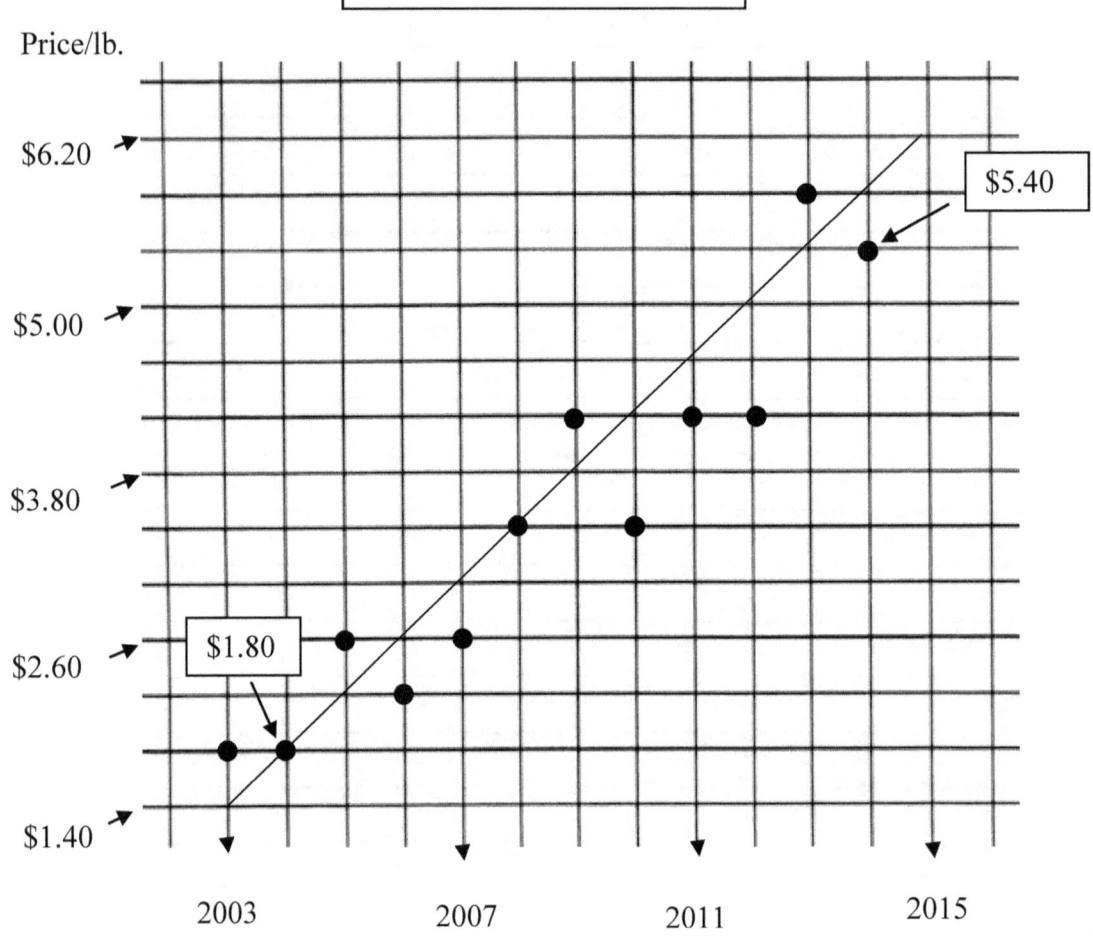

$$Percentage\ Increase = \frac{Change\ in\ Price}{Original\ Price} = \frac{5.40 - 1.80}{1.80} = 2 = 200\%$$

Answer: (A) 200%

56. Solve for x in the inequality 5|2x − 12| ≤ 7.

(A) $x \leq 6.7$ (B) $5.3 \leq x$ (C) $-4.6 \leq x \leq 2.7$ (D) $5.3 \leq x \leq 6.7$ (E) $-6.7 \leq x \leq -5.3$

Set the inequality between −7 and +7, while removing the absolute value symbol.

57. Given the Universal set $U = \{r, s, t, u\}$ and the operations Δ and θ as defined by the tables shown, find $(u \, \Delta \, s) \, \theta \, u$.

(A) t   (B) u   (C) r   (D) s   (E) 1

Δ	r	s	t	u
r	t	u	r	s
s	u	r	s	t
t	r	s	t	u
u	s	t	u	r

θ	r	s	t	u
r	s	t	u	r
s	t	u	r	s
t	u	r	s	t
u	r	s	t	u

Find $u \, \Delta \, s$ in the first table. Use that answer to find $(u \, \Delta \, s) \, \theta \, u$ in the second table.

56. Solve for x in the inequality $5|2x - 12| \leq 7$.

(A) $x \leq 6.7$ (B) $5.3 \leq x$ (C) $-4.6 \leq x \leq 2.7$ (D) $5.3 \leq x \leq 6.7$ (E) $-6.7 \leq x \leq -5.3$

$$5|2x - 12| \leq 7$$
$$-7 \leq 5(2x - 12) \leq 7$$
$$-7 \leq 10x - 60 \leq 7$$
$$53 \leq 10x \leq 67$$
$$5.3 \leq x \leq 6.7$$

Answer: (D) $5.3 \leq x \leq 6.7$

57. Given the Universal set $U = \{r, s, t, u\}$ and the operations Δ and θ as defined by the tables shown, find $(u \Delta s) \theta u$.

(A) t (B) u (C) r (D) s (E) 1

Δ	r	s	t	u
r	t	u	r	s
s	u	r	s	t
t	r	s	t	u
u	s	t	u	r

θ	r	s	t	u
r	s	t	u	r
s	t	u	r	s
t	u	r	s	t
u	r	s	t	u

First calculate $u \Delta s$, using the Δ-table:

Δ	r	s	t	u
r	t	u	r	s
s	u	r	s	t
t	r	s	t	u
u	s	t	u	r

Going across row u and down column s yields the entry t.

So, $(u \Delta s) \theta u$ becomes $t \theta u$. Using the θ-table, we get

θ	r	s	t	u
r	s	t	u	r
s	t	u	r	s
t	u	r	s	t
u	r	s	t	u

Going across row t and down column u yields the entry t.

Answer: (A) t

58. If each of the marks on the x-axis represents one unit, which equation represents the parabola?

(A) $y = x^2 - 4x - 5$ (B) $y = x^2 + x - 3$ (C) $y = x^2 + x - 3$ (D) $y = x^2 + 2x - 1 = 0$
(E) $y = 2x^2 - x - 2$

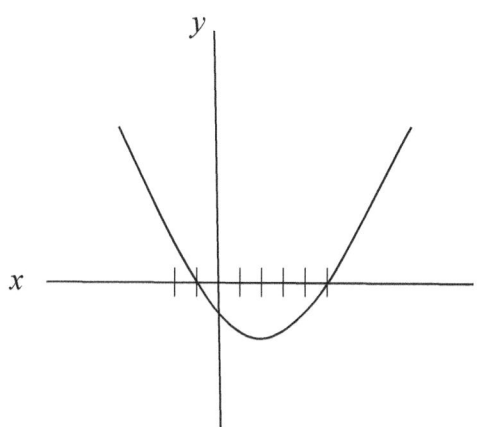

Each x-coordinate where the parabola crosses the x-axis is a root of the equation $y = 0$.

59. What is the cot of 330°?

(A) $\sqrt{3}$ (B) $\dfrac{1}{2}$ (C) $\dfrac{\sqrt{3}}{2}$ (D) $\dfrac{1}{3}$ (E) $-\sqrt{3}$

$$\cot\theta = \dfrac{1}{\tan\theta}$$

58. If each of the marks on the x-axis represents one unit, which equation represents the parabola?

(A) $y = x^2 - 4x - 5$ (B) $y = x^2 + x - 3$ (C) $y = x^2 + x - 3$ (D) $y = x^2 + 2x - 1 = 0$
(E) $y = 2x^2 - x - 2$

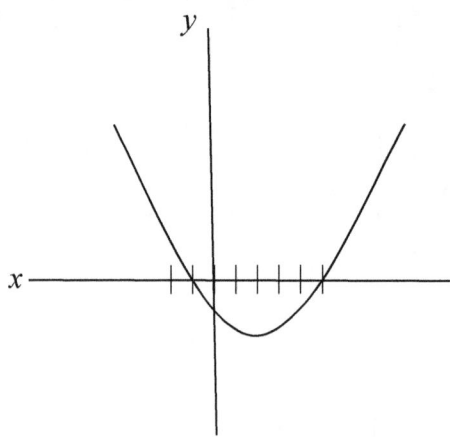

The graph crosses the x-axis at 5. Hence, $x = 5$ is a root. That means $x - 5$ is a factor of the equation $y = 0$. Additionally, the graph crosses the x-axis at -1. Hence, $x = -1$ is a root. That means $x - (-1) = x + 1$ is also a factor of the equation $y = 0$:

$$(x - 5)(x + 1) = 0$$
$$x^2 - 4x - 5 = 0$$
$$y = x^2 - 4x - 5$$

Answer: (A) $y = x^2 - 4x - 5$

59. What is the cot of 330°?

(A) $\sqrt{3}$ (B) $\dfrac{1}{2}$ (C) $\dfrac{\sqrt{3}}{2}$ (D) $\dfrac{1}{3}$ (E) $-\sqrt{3}$

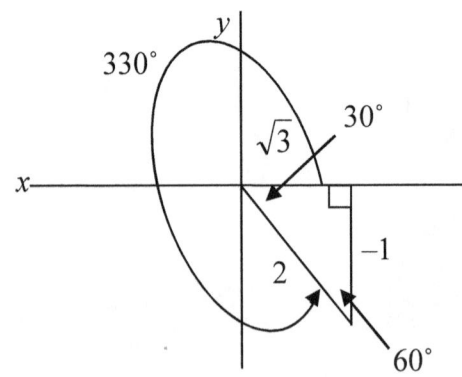

$$\cot 330° = \frac{1}{\tan 330°}$$
$$= \frac{1}{-1/\sqrt{3}} = -\sqrt{3}$$

Answer: (E) $-\sqrt{3}$

60. Point O(5, 6) is the center of a circle on the coordinate plane. If point A(1, 3) is located on the circumference of the circle, what is the equation of the circle?

(A) $(x + 6)^2 + (y - 5)^2 = 36$ (B) $(x - 6)^2 + (y + 3)^2 = 25$ (C) $x^2 + y^2 = 36$
(D) $(x - 5)^2 + (y - 6)^2 = 25$ (E) $(x - 3)^2 + (y - 4)^2 = 25$

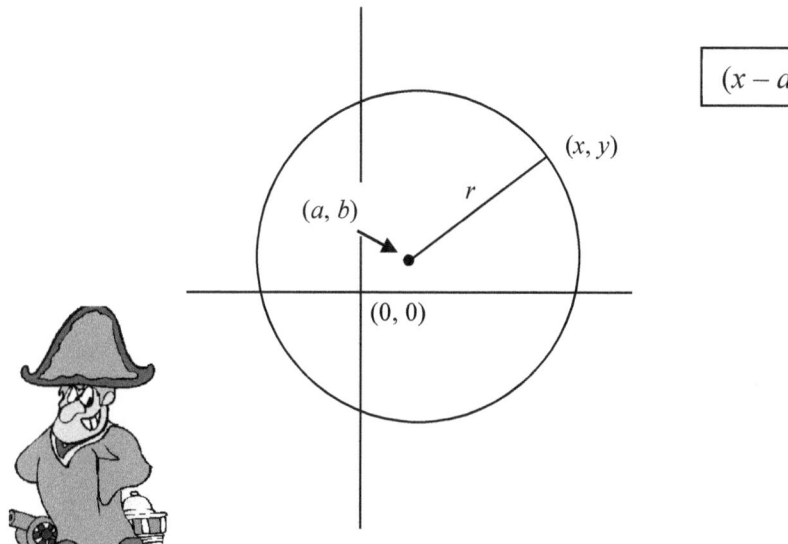

60. Point O(5, 6) is the center of a circle on the coordinate plane. If point A(1, 3) is located on the circumference of the circle, what is the equation of the circle?

(A) $(x + 6)^2 + (y - 5)^2 = 36$
(B) $(x - 6)^2 + (y + 3)^2 = 25$
(C) $x^2 + y^2 = 36$
(D) $(x - 5)^2 + (y - 6)^2 = 25$
(E) $(x - 3)^2 + (y - 4)^2 = 25$

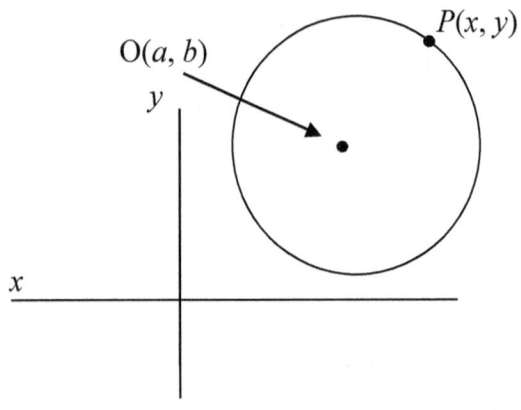

General form for the equation of a circle is $(x - a)^2 + (y - b)^2 = \text{radius}^2$, where a and b are coordinates of the center of the circle and x and y are the coordinates of a point on the circumference.

$(x - 5)^2 + (y - 6)^2 = 5^2$

Answer: (D) $(x - 5)^2 + (y - 6)^2 = 25$

Answer Key to Test 2

1. B	21. E	41 D
2. E	22. E	42. A
3. B	23. D	43. E
4. D	24. A	44. C
5. C	25. A	45. E
6 A	26. B	46. B
7. B	27. C	47. D
8. E	28. D	48. E
9. D	29. D	49. B
10. C	30. A	50. C
11. E	31. C	51. A
12. C	32. C	52. D
13. A	33. A	53. B
14. B	34. E	54. E
15. A	35. D	55. A
16. D	36. D	56. D
17. B	37. E	57. A
18. B	38. C	58. A
19. A	39. B	59. E
20. B	40. B	60. D

Test 3

1. For positive integers a and b, $a \phi b = \dfrac{a^2}{b-1} + 6$. Using this definition, simplify $(3 \phi 4) \phi 2$.

(A) 87 (B) 64 (C) 72 (D) 82 (E) 56

2. The relative costs of manufacturing a motorcycle are pictured in the circle graph shown. If it costs $1,800 to manufacture a motorcycle, how much does it cost to assemble a tire?

(A) $350 (B) $400 (C) $450 (D) $250 (E) $300

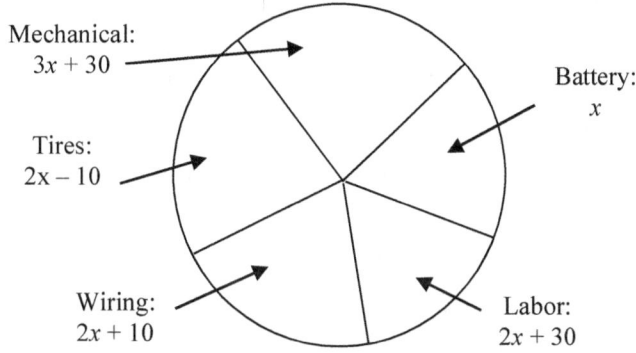

Cost of Manufacturing a Motorcycle
Mechanical: $3x + 30$
Battery: x
Tires: $2x - 10$
Wiring: $2x + 10$
Labor: $2x + 30$

3. If $1 = b^{2x-2}$, find the value of x.

(A) 6 (B) 1 (C) 3 (D) 2 (E) 4

4. A circle of radius 2 inches and a square 4 inches on each side are on the same plane. What is the maximum number of points at which they can touch?

(A) 1 (B) 4 (C) 3 (D) 2 (E) 5

5. If $a^2 - a = 30$, find the positive value of a.

(A) 2 (B) 4 (C) 3 (D) 5 (E) 6

6. The letters *r*, *s* and *t* represent positive digits and *r/s* is less than one. The number represented by *t* is midway between *r* and *s*. What is *t* equal to?

(A) $r + \dfrac{s-r}{2}$ (B) $r + \dfrac{s}{2}$ (C) $r + s - \dfrac{r}{2}$ (D) $\dfrac{s+r}{2}$ (E) $r + \dfrac{s+r}{2}$

7. Simplify

$$\dfrac{6x^{3/2} + 2x^{1/2}}{2/x^{1/2}}$$

(A) $4x^2 + x$ (B) $2x^3 - 2x$ (C) $3x^2 + x$ (D) $4x^2 - x$ (E) $3x^2 + 2x$

8. The sum of 3 consecutive odd integers is 1 less than the sum of the 2 consecutive even integers immediately following the last odd integer. Find the first even integer following the third odd integer.

(A) 10   (B) 12   (C) 8   (D) 14   (E) 16

9. A rocket is shot upward. What is its initial velocity if after 10 seconds it reaches a height of 48,400 feet? Use the formula $h = vt - \dfrac{1}{2}gt^2$, where h = height of the rocket, v = initial launch velocity, t = time (in seconds) and g = the downwards pull of gravity = 32.

(A) 3,600 ft/sec   (B) 5,000 ft/sec   (C) 2,000 ft/sec   (D) 6,000 ft/sec
(E) 5,500 ft/sec

10. Simplify

$$\dfrac{(4r)^{-2}}{(12r)^{-1}}$$

(A) $\dfrac{4}{3r^2}$ (B) $3r^2$ (C) $\dfrac{3}{4r}$ (D) $\dfrac{1}{3r}$ (E) $3r^{-1}$

11. If $\tan x = -1$ and $0° \leq x \leq 180°$, find x.

(A) 110°   (B) 150°   (C) 135°   (D) 90°   (E) 125°

12. Given: All elements in set A are elements in set B. Assuming that B is larger than A, which conclusions are true?

(A) Some A are not B. (B) Some B are A. (C) No A are B.
(D) No B are A. (E) Some not B are A.

13. The center of a circle is located at (3, 5). If the circumference of the circle is 44, find the equation representing this circle. Let π = 22/7.

(A) $(x + 3)^2 + (y + 5)^2 = 16$ (B) $x^2 + y^2 = 49$ (C) $(x - 3)^2 + (y - 5)^2 = 16$
(D) $x^2 + y^2 = 16$ (E) $(x - 3)^2 + (y - 5)^2 = 49$

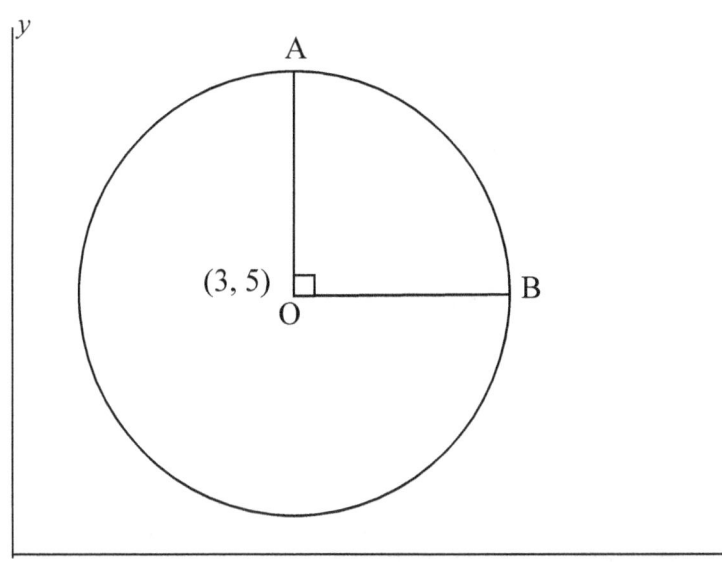

14. In the diagram shown, AB = 2CD, AD = 8.1 and CD = .9. Find the ratio of BC : AC.

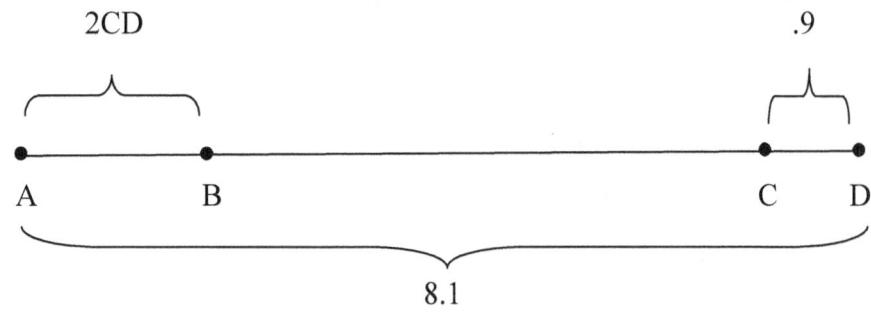

(A) 3/4 (B) 4/5 (C) 2/3 (D) 4/7 (E) 5/6

ACT Math Personal Tutor

15. If $f(x) = 5x - 2$ and $f(g(x)) = 10x$, find $g(x)$.

(A) $5x + 2$ (B) $2x + 2/5$ (C) $3x - 2/3$ (D) $x - 1/3$ (E) $3x + 2$

16. In Mandy's Coffee Shop, for every a cups of coffee sold, $\sqrt{b}$ cups of tea are sold. Using the same ratio, how many cups of tea are sold when c cups of coffee are sold?

(A) $a\dfrac{\sqrt{b}}{c}$ (B) $\dfrac{\sqrt{b}}{ac}$ (C) $c\dfrac{\sqrt{b}}{a}$ (D) $\dfrac{ac}{\sqrt{b}}$ (E) $ac\sqrt{b}$

17. Find the equation of the line connecting points A(2, 3) and B(4, 9).

(A) $y = 2x - 4$ (B) $y = 2x + 3$ (C) $y = 2x - 4$ (D) $y = 3x - 3$ (E) $y = 3x + 1$

18. What is the approximate value of $\dfrac{2.7834}{7} \times 526$?

(A) 310 (B) 180 (C) 225 (D) 380 (E) 450

19. If the temperature (T) of a star falls by 2% each million years and the temperature now is 1,000,000°, derive the formula for the temperature 10 million years from now.

(A) $1,000,000(.98)^{10}$ (B) $.02(1,000,000)^{10}$ (C) $1,000,000^9(.98)$
(D) $[(.02)(1,000,000)]^{10}$ (E) $1,000,000(.98)^9$

20. If the diameters of each of the smaller circles is 1/3 the diameter of the large circle, find the area in between the large circle and the smaller circles.

(A) $\dfrac{\pi r^2}{8}$ (B) $\dfrac{2\pi r^2}{9}$ (C) $\dfrac{\pi r^2}{27}$ (D) $\dfrac{5\pi r^2}{9}$ (E) $\dfrac{5\pi r^2}{18}$

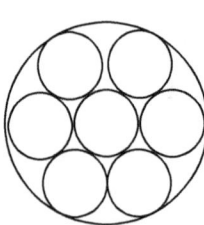

Test 3

21. Find the equation of the perpendicular bisector of the line connecting the points A(–2, 0) and B(2, –8).

(A) $y = \frac{1}{2}x - 4$ (B) $y = 3x - 2$ (C) $y = \frac{1}{2}x + 3$ (D) $y = 2x + 4$ (E) $y = \frac{1}{2}x + 2$

22. Multiply the two complex numbers $(4 + 2i)$ and $(3 - 5i)$.

(A) $6 - 3i$ (B) $14 + 12i$ (C) $16 + 2i$ (D) $22 - 14i$ (E) $16 - 5i$

23. Solve for x in the equation

$$\frac{4}{\sqrt{2x+5}} = \frac{\sqrt{2x+5}}{x}$$

(A) 2 (B) $2\frac{1}{2}$ (C) $-3\frac{1}{2}$ (D) -3 (E) $-3\frac{1}{3}$

24. An inspector checks b number of batteries every 10 minutes. If $.02c$ batteries fail every 1/2 hour, what fraction pass after 6 hours?

(A) $\frac{40b - .16c}{24b}$ (B) $\frac{2.6b - 10c}{36c}$ (C) $\frac{36b - .24c}{36b}$ (D) $\frac{16c - .04b}{24b - 2c}$ (E) $\frac{.3b - .24c}{.8b + c}$

25. Find the distance between A(10, 11) and B(2, 5) on the xy-plane.

(A) 10 (B) 8 (C) $\sqrt{46}$ (D) $\sqrt{51}$ (E) 9

26. Given the statement "If $\sqrt{x+4} \geq 3$, then $x > 5$," which one of the following statements is a counterexample.

(A) $x < 5$ (B) $x = 2$ (C) $4 < x < 5$ (D) $x \leq 3$ (E) $x = 5$

27. Rationalize the denominator:

$$\frac{6 - 5i}{3 + 4i}$$

(A) $\frac{4 + 6i}{23}$ (B) $\frac{-2 - 39i}{25}$ (C) $\frac{5 - 6i}{14}$ (D) $\frac{2 + 3i}{26}$ (E) $\frac{7 + 8i}{18}$

ACT Math Personal Tutor

28. Find the vertical difference between the graphs of $y = 2x^2 - x + 4$ and $y = x^2 + 2x + 3$ at $x = -1$.

(A) 3   (B) 4   (C) 5   (D) 2   (E) 6

29. If the 4th term of an arithmetic sequence is –9 and the first term is –3, find the median of the first 6 terms.

(A) –7   (B) –8   (C) –9   (D) –10   (E) 6

30. Determine the range of the function $y = 3 \cos 4x$.

(A) $-3 \leq y \leq 3$ (B) $-4 \leq y \leq 4$ (C) $-3 \leq y \leq 0$ (D) $-4 \leq y \leq 0$ (E) $-3 \leq y \leq 4$

31. Miller's Department store sells a television set at a 20% discount. Jackson's, another department store, sells the same set for $25 more than Miller's discounted price. If Jackson's sells the set for $505, how much was the original non-discounted price?

(A) $560   (B) $580   (C) $600   (D) $620   (E) $490

32. Given the following two equations, solve for s in terms of v.

$$(1) \ r + 2s = t$$
$$(2) \ t - r = v$$

(A) $s = \dfrac{v}{3}$ (B) $s = \dfrac{2v}{3}$ (C) $s = \dfrac{v}{4}$ (D) $s = \dfrac{v}{2}$ (E) $s = \dfrac{3v}{2}$

33. Reflect $\triangle ABC$ with vertices A(3, 1), B(5, 1) and C(5, 3) along the line $y = x$ into its image A'B'C'. What are the co-ordinates of point B'?

(A) (1, 5)   (B) (2, 5)   (C) (4, 5)   (D) (1, 4)   (E) (2, 4)

34. A poll of people's reading preferences was taken and the results (expressed in degree measure) are pictured in the circle shown. If 1,800 people enjoy fiction, how many people in the poll enjoy history?

(A) 3,000

(B) 2,600

(C) 3,200

(D) 2,700

(E) 2,900

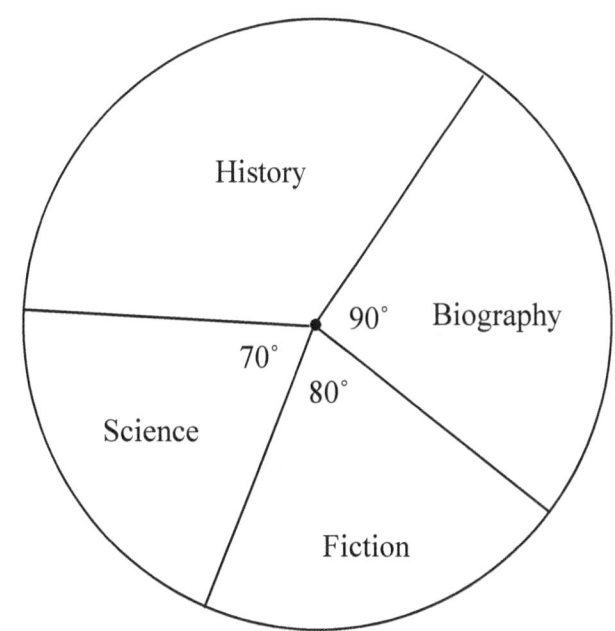

35. Given the two equations $y + 2x = 0$ and $3^{xy} = \frac{1}{81}$, find the positive value of x.

(A) 2 (B) $\sqrt{2}$ (C) $\sqrt{3}$ (D) 3 (E) 1/3

36. If $b = \sqrt[3]{c}$ and $8 \leq c \leq 125$ and $a = \frac{b}{c}$, find the greatest possible value for a.

(A) 5/8 (B) 1/2 (C) 7/8 (D) 1/4 (E) 3/4

37. If the volume of a cube is 125 cubic inches, what is its total surface area?

(A) 120 sq in (B) 225 sq in (C) 150 sq in (D) 180 sq in (E) 210 sq in

38. The table below indicates values of the function $f(x)$ for given values of x.

x	1	2	3	4
$f(x)$	2	5	10	17

Which function is most likely represented by this table?

(A) $3x^2 - 1$ (B) $x^2 + 2$ (C) $2x^2 + 3$ (D) $x^2 - x + 1$ (E) $x^2 + 1$

39. Jack goes to the Nuart Theater every 2nd day at noon, Myra goes to the same theater every 3rd day at noon, and Jose goes to the same theater every 4th day at noon. If they all go to the Nuart on the same day, when is the earliest day they will they all meet again at the theater?

(A) 8 days (B) 13 days (C) 10 days (D) 11 days (E) 9 days

40. If the average 18-year old would reduce her daily intake of sugar by 1/5, she would still consume 10 more grams of sugar a day than the average 26-year old. If the average 26-year old consumes 70 grams of sugar per day, how many grams of sugar does the average 18-year old consume?

(A) 300 (B) 900 (C) 400 (D) 100 (E) 80

41. The electrical resistance of a wire, r, varies directly as its length, l, and inversely as the square of its diameter, d. The resistance of a wire 400 inches in length with a diameter of 1/4 of an inch is 32 ohms. Find the resistance of a wire 200 inches in length with a diameter of 1/8 of an inch.

(A) 20 ohms (B) 64 ohms (C) 44 ohms (D) 32 ohms (E) 28 ohms

42. If $|x| y^{2/3} = 80$ and $x = -5$, what is the value of y?

(A) 48 (B) 32 (C) 16 (D) 64 (E) 80

43. The length of a rectangle is 4 inches more than its width. If the length is doubled and the width is reduced by 2 inches, the area of the new rectangle is 20 square inches greater than the area of the original rectangle. Find the length of the original rectangle.

(A) 10 (B) 12 (C) 9 (D) 8 (E) 14

44. Find the 48th digit of the repeating decimal 0.26734....

(A) 2   (B) 6   (C) 7   (D) 3   (E) 4

45. An equilateral triangle is inscribed in a circle of radius 4 as shown. Find the shaded area. Leave your answer in terms of π and radicals.

(A) $\dfrac{3\pi + 4\sqrt{3}}{4}$ (B) $\dfrac{16\pi - 12\sqrt{3}}{3}$ (C) $\dfrac{12\pi + 2\sqrt{2}}{3}$ (D) $\dfrac{12\pi - 2\sqrt{3}}{4}$ (E) $\dfrac{16\pi - 4\sqrt{3}}{4}$

Note: Area $\Delta = \dfrac{1}{2} a \cdot b \sin C$, where a and b are sides of the Δ and C is the included angle.

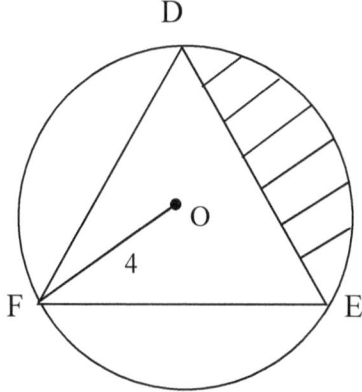

46. Find the value of $x + y$ in the matrix addition problem shown.

$$\begin{bmatrix} 2 & 4 \\ 3 & y \end{bmatrix} + \begin{bmatrix} 5 & x \\ 2 & 7 \end{bmatrix} = \begin{bmatrix} 7 & 10 \\ 5 & 6 \end{bmatrix}$$

(A) –1   (B) 4   (C) 5   (D) 3   (E) –2

47. A ball is dropped from a height of 24 feet. If it continues to bounce up 1/2 the dropped height, how many times does it hit the ground when the height after the last bounce measures 3/4 of a foot?

(A) 1 bounce   (B) 2 bounces   (C) 3 bounces   (D) 4 bounces   (E) 5 bounces

48. Which of the choices has the same roots as $f(x) = 2x^2 - 2x - 12$?

(A) $x^2 - x - 6$ (B) $x^2 + 6x - 2$ (C) $3x^2 - 2x - 4$ (D) $x^2 + 4x - 3$ (E) $2x^2 + 3x - 5$

49. There are 4 positions available for the chess team. If 9 students apply, in how many ways can the positions be filled?

(A) 2,016 (B) 4,268 (C) 3,024 (D) 4,561 (E) 2,884

50. We want to get 100 ounces of a 44% solution of alcohol by mixing an 80% solution of alcohol with a 20% solution of alcohol. How many ounces of the 20% solution should we include?

(A) 20 ounces (B) 50 ounces (C) 30 ounces (D) 40 ounces (E) 60 ounces

51. Simplify

$$\frac{2x+8}{2x^2+6x-8}$$

(A) $\dfrac{2x}{x-4}$ (B) $\dfrac{1}{x-1}$ (C) $\dfrac{3x}{2x+1}$ (D) $\dfrac{2}{3x-2}$ (E) $\dfrac{4x}{2x+1}$

52. Determine the axis of symmetry of the graph represented by the function $f(x) = (2x + 1)(x + 3)$.

(A) –3/4 (B) –7/4 (C) 5/8 (D) 9/7 (E) –8/5

53. A particular type of plant has been dying at the rate of 3% per year. If there are 332,520 plants at the current time, how many plants were there 6 years ago? Use the formula $A = i(1 - r)^n$, where A = the final number, i = the initial population, r = % decrease per time period and n = the number of time periods. Round off to the nearest whole number

(A) 375,420 (B) 401,426 (C) 399,197 (D) 385,176 (E) 380,526

54. The average of three whole numbers, a, b and c is 24. If b is 2 more than a and c is 14 more than b, what is the value of a?

(A) 6 (B) 8 (C) 18 (D) 22 (E) 16

55. If $f(x) = 2x^2 + 3$ and $g(x) = x - 4$, find the value of $g\left(f(\sqrt{2})\right)$.

(A) 3 (B) 4 (C) 5 (D) 6 (E) 7

56. Identify one point where the graph of the line $f(x) = 0$ cuts through the graph of $g(x) = x^3 - 2x$?

(A) (2, 0)   (B) (0, $\sqrt{2}$)   (C) (0, 2)   (D) ($\sqrt{2}$, 0)   (E) ($\sqrt{3}$, 0)

57. In a survey of the ice cream preferences of 80 persons, 18 liked vanilla alone, 11 liked chocolate alone, and 15 liked a mix of chocolate and strawberry. Six liked strawberry alone, and the rest liked a mix of strawberry and vanilla. How many liked a mix of vanilla and strawberry?

(A) 31   (B) 36   (C) 19   (D) 23   (E) 30

58. Which of the following graphs represents the equation $y = \dfrac{1}{x^2}, x > 0$?

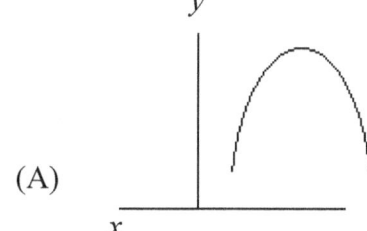

(A)   (B)   (C)

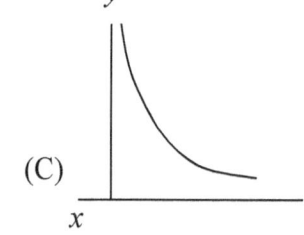

(D)   (E)

59. Find the value of $a - b$ in the matrix subtraction problem below.

(A) 3   (B) –4   (C) 6   (D) –1   (E) –2

$$\begin{bmatrix} 2 & 4 \\ 5 & -1 \end{bmatrix} - \begin{bmatrix} a & 3 \\ 6 & 2 \end{bmatrix} = \begin{bmatrix} 6 & 7 \\ -1 & b \end{bmatrix}$$

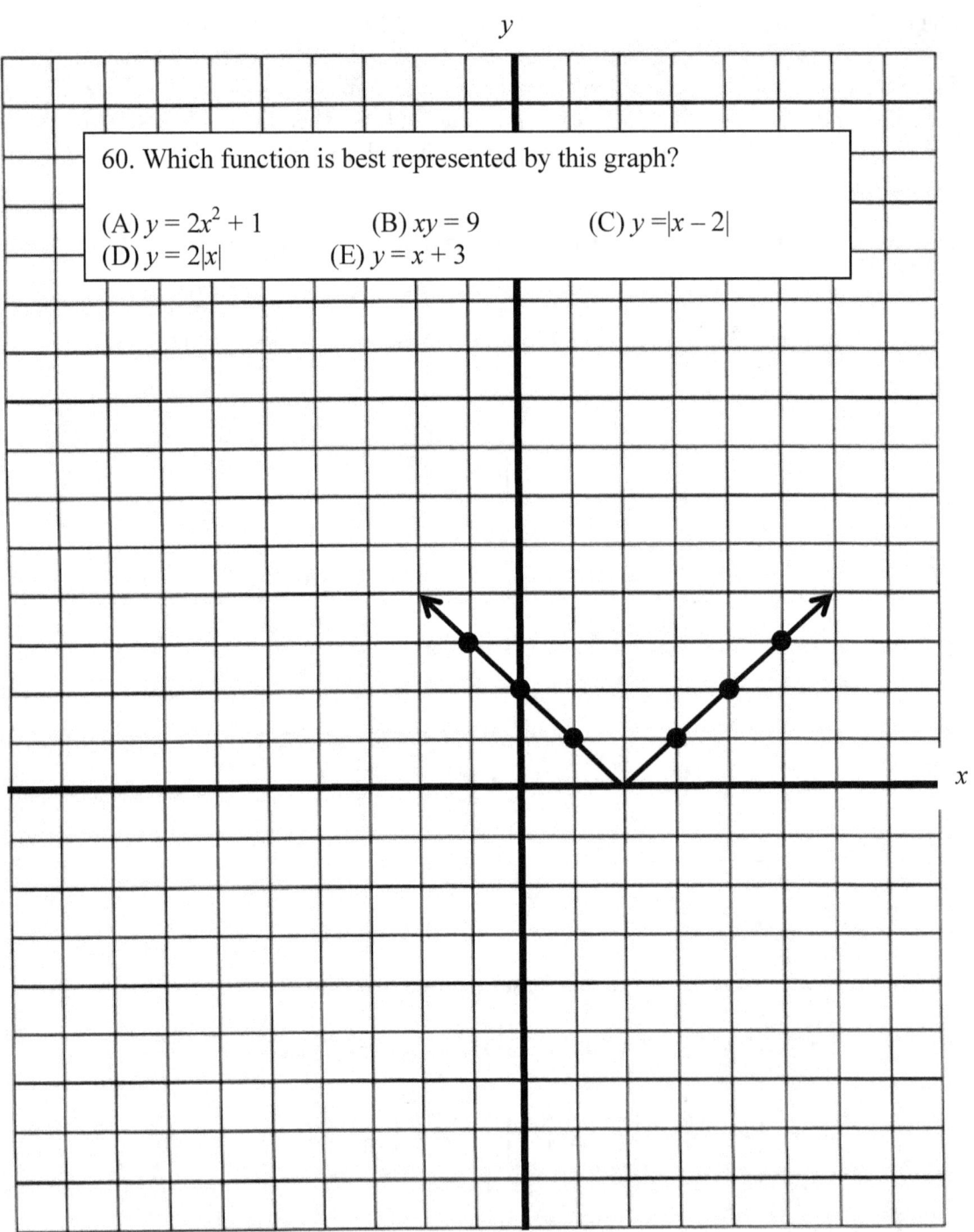

60. Which function is best represented by this graph?

(A) $y = 2x^2 + 1$ (B) $xy = 9$ (C) $y = |x - 2|$
(D) $y = 2|x|$ (E) $y = x + 3$

Answers to Test 3

1. For positive integers a and b, $a \phi b = \dfrac{a^2}{b-1} + 6$. Using this definition, simplify $(3 \phi 4) \phi 2$.

(A) 87   (B) 64   (C) 72   (D) 82   (E) 56

$$3 \phi 4 = \frac{3^2}{4-1} + 6 = \frac{9}{3} + 6 = 3 + 6 = 9$$

So, $(3 \phi 4) \phi 2$ becomes $9 \phi 2$.

$$9 \phi 2 = \frac{9^2}{2-1} + 6 = \frac{81}{1} + 6 = 87$$

Answer: (A) 87

2. The relative costs of manufacturing a motorcycle are pictured in the circle graph shown. If it costs $1,800 to manufacture a motorcycle, how much does it cost to assemble a tire?

(A) $350   (B) $400   (C) $450   (D) $250   (E) $300

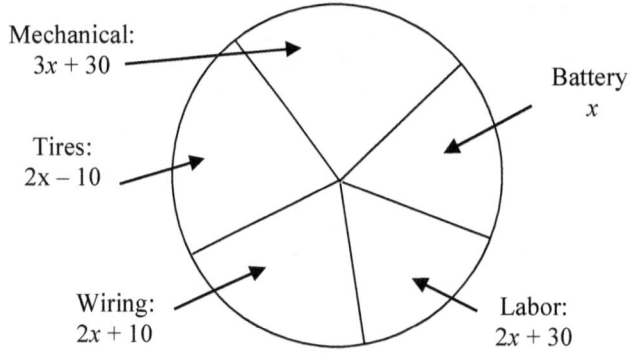

The total costs makeup 360° of the circle:

$x + (2x + 30) + (2x + 10) + (2x - 10) + (3x + 30) = 360°$
$10x + 60 = 360°$
$10x = 300°$
$x = 30°$

Tires: $2x - 10 = 2(30°) - 10° = 50°$

$\dfrac{50°}{360°} \cdot \$1800 = \$250$

Answer: (D) $250

Test 3

3. If $1 = b^{2x-2}$, find the value of x.

(A) 6 (B) 1 (C) 3 (D) 2 (E) 4

$$1 = b^{2x-2}$$

To solve an exponential equation, we first try to equate the exponents of equal bases.*
The base of the right side of the equation is b, and there is no base on the left side of the equation. However, by definition $b^0 = 1$. So, we get

$$b^0 = b^{2x-2}$$

Now, equating the exponents of the equal bases (b) gives

$$0 = 2x - 2$$

$$2x = 2$$

$$x = 1$$

Answer (B) 1

---

* If that cannot be done, then we use logs.

4. A circle of radius 2 inches and a square 4 inches on each side are on the same plane. What is the maximum number of points at which they can touch?

(A) 1   (B) 4   (C) 3   (D) 2   (E) 5

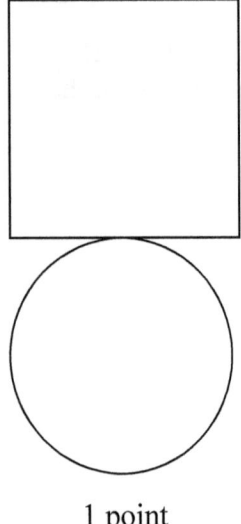
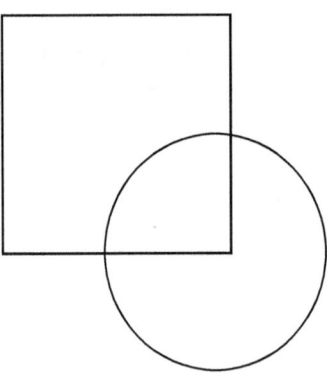
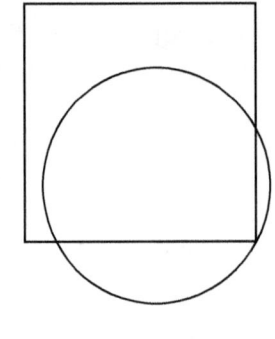
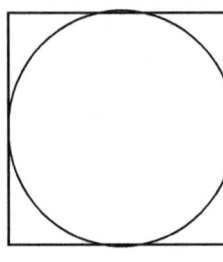

1 point   2 points   3 points   4 points

Answer: (B) 4

5. If $a^2 - a = 30$, find the positive value of a.

(A) 2   (B) 4   (C) 3   (D) 5   (E) 6

$$a^2 - a = 30$$
$$a^2 - a - 30 = 0$$
$$(a - 6)(a + 5) = 0$$
$$a - 6 = 0 \text{ or } a + 5 = 0$$
$$a = 6 \text{ or } a = -5$$

Answer: (E) 6

6. The letters *r*, *s* and *t* represent positive digits and *r/s* is less than one. The number represented by *t* is midway between *r* and *s*. What is *t* equal to?

(A) $r + \dfrac{s-r}{2}$ (B) $r + \dfrac{s}{2}$ (C) $r + s - \dfrac{r}{2}$ (D) $\dfrac{s+r}{2}$ (E) $r + \dfrac{s+r}{2}$

Since *r/s* < 1, we know that *r* < *s*.

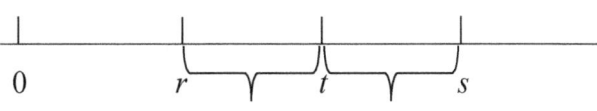

The distance between *s* and *r* is *s* − *r*. Since *t* is the midpoint between *r* and *s*, the distance between *t* and *r* is

$$\dfrac{s-r}{2}$$

Now, the distance from 0 to *t* equals (the distance from 0 to *r*) + (the distance from *r* to *t*).

So, $t = (r - 0) + \left(\dfrac{s-r}{2}\right) = r + \dfrac{s-r}{2}$

Answer: (A) $r + \dfrac{s-r}{2}$

7. Simplify

$$\dfrac{6x^{3/2} + 2x^{1/2}}{2/x^{1/2}}$$

(A) $4x^2 + x$ (B) $2x^3 - 2x$ (C) $3x^2 + x$ (D) $4x^2 - x$ (E) $3x^2 + 2x$

$$\dfrac{6x^{3/2} + 2x^{1/2}}{2/x^{1/2}} =$$

$$\dfrac{6x^{3/2} + 2x^{1/2}}{1} \cdot \dfrac{x^{1/2}}{2} =$$

$$\dfrac{2(3x^{3/2} + x^{1/2})}{1} \cdot \dfrac{x^{1/2}}{2} =$$

$$(3x^{3/2} + x^{1/2}) \cdot x^{1/2} =$$

$$3x^{3/2} \cdot x^{1/2} + x^{1/2} \cdot x^{1/2} =$$

$$3x^{\frac{3}{2}+\frac{1}{2}} + x^{\frac{1}{2}+\frac{1}{2}} =$$

$$3x^{\frac{4}{2}} + x^{\frac{2}{2}} =$$

$$3x^2 + x$$

Answer: (C) $3x^2 + x$

8. The sum of 3 consecutive odd integers is 1 less than the sum of the 2 consecutive even integers immediately following the last odd integer. Find the first even integer following the third odd integer.

(A) 10 (B) 12 (C) 8 (D) 14 (E) 16

Let x be the first odd integer.
Then, $x + 2$ is the second odd integer.
And $x + 4$ is the third odd integer.
So, $(x + 4) + 1$ is the first even integer following the third odd integer.
And $(x + 4 + 1) + 2$ is the second even integer following the third odd integer.

Now, translating the clause "the sum of 3 consecutive odd integers is 1 less than the sum of the 2 consecutive even integers immediately following the last odd integer" into an equation yields

$$x + (x + 2) + (x + 4) = (x + 4 + 1) + (x + 4 + 1 + 2) - 1$$
$$3x + 6 = 2x + 11$$
$$x = 5$$

First odd integer: $x = 5$
Second odd integer: $x + 2 = 5 + 2 = 7$
Third odd integer: $x + 4 = 5 + 4 = 9$
First even integer following the third odd integer: $(x + 4) + 1 = 9 + 1 = 10$

Answer: (A) 10

9. A rocket is shot upward. What is its initial velocity if after 10 seconds it reaches a height of 48,400 feet? Use the formula $h = vt - \frac{1}{2}gt^2$, where h = height of the rocket, v = initial launch velocity, t = time (in seconds) and g = the downwards pull of gravity = 32.

(A) 3,600 ft/sec (B) 5,000 ft/sec (C) 2,000 ft/sec (D) 6,000 ft/sec
(E) 5,500 ft/sec

$$h = vt - \frac{1}{2}gt^2$$

$h = 48,400, t = 10, g = 32$: $48,400 = v(10) - \frac{1}{2}(32)(10)^2$
$$48,400 = 10v - 1,600$$
$$50,000 = 10v$$
$$v = 5,000$$

Answer: (B) 5,000 ft/sec

10. Simplify

$$\frac{(4r)^{-2}}{(12r)^{-1}}$$

(A) $\frac{4}{3r^2}$ (B) $3r^2$ (C) $\frac{3}{4r}$ (D) $\frac{1}{3r}$ (E) $3r^{-1}$

$$\frac{(4r)^{-2}}{(12r)^{-1}} = \frac{(12r)^{+1}}{(4r)^2} = \frac{12r}{16r^2} = \frac{3}{4r}$$

Answer: (C) $\frac{3}{4r}$

11. If $\tan x = -1$ and $0° \leq x \leq 180°$, find x.

(A) 110° (B) 150° (C) 135° (D) 90° (E) 125°

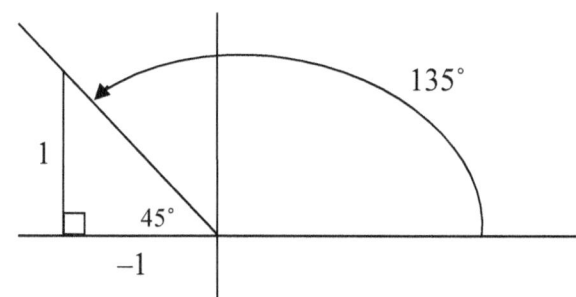

Recall that $\tan x = +1$ when $x = 45°$, so the reference angle is 45°. Now, angle x is in Quadrant II because all trig functions are positive in Quadrant I and we are given that $\tan x$ is negative and that $0° \leq x \leq 180°$. Hence, $x = 180 - 45° = 135°$ as displayed in the figure.

Answer: (C) 135°

12. Given: All elements in set A are elements in set B. Assuming that B is larger than A, which conclusions are true?

(A) Some A are not B. (B) Some A are B. (C) No A are B.
(D) No B are A. (E) Some not B are A.

Let's draw a Venn diagram. Since "all elements in set A are elements in set B," the circle representing set A is entirely within the circle representing set B:

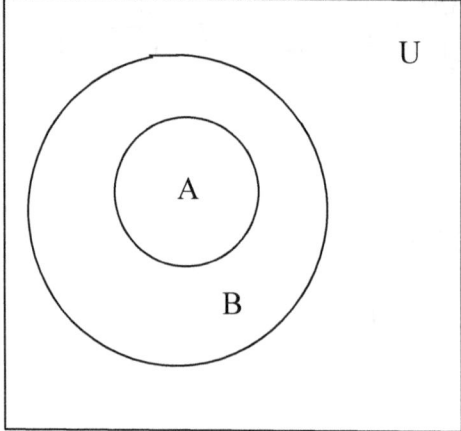

From the diagram, we get

Some A are not B.	False
Some A are B.	True (In fact, all A are B.)
No A are B.	False
No B are A.	False
Some not B are A.	False

Answer: (B) Some A are B.

13. The center of a circle is located at (3, 5). If the circumference of the circle is 44, find the equation representing this circle. Let $\pi = 22/7$.

(A) $(x + 3)^2 + (y + 5)^2 = 16$ (B) $x^2 + y^2 = 49$ (C) $(x - 3)^2 + (y - 5)^2 = 16$
(D) $x^2 + y^2 = 16$ (E) $(x - 3)^2 + (y - 5)^2 = 49$

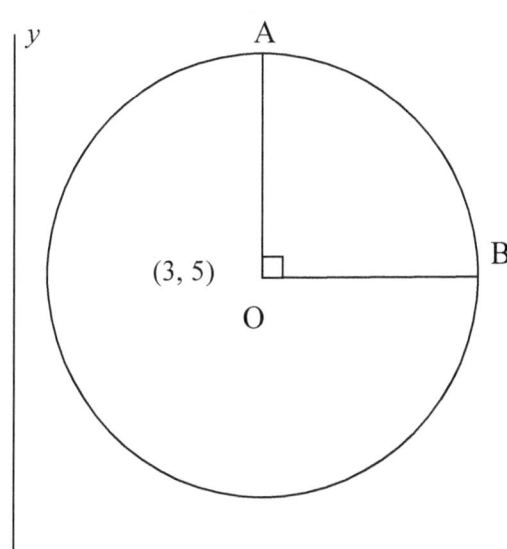

Use the formula $(x - a)^2 + (y - b)^2 = r^2$, where a and b are coordinates of the center of the circle and x and y are coordinates of a point on the circumference.

Let the circumference be C and the radius be r.

$$C = 2\pi r$$

$C = 44$, $\pi = 22/7$:

$$44 = 2 \cdot \frac{22}{7} r$$

$$44 = \frac{44}{7} r$$

$$r = 7$$

Since the center of the circle is at (3, 5) and the radius is 7, the formula for the equation of a circle gives

$$(x - 3)^2 + (y - 5)^2 = 7^2$$

Answer: (E) $(x - 3)^2 + (y - 5)^2 = 49$

14. In the diagram shown, AB = 2CD, AD = 8.1 and CD = .9. Find the ratio of BC : AC.

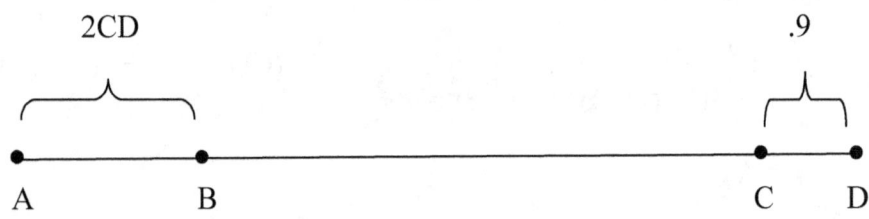

(A) 3/4   (B) 4/5   (C) 2/3   (D) 4/7   (E) 5/6

$$AB = 2 \cdot CD = 2 \cdot .9 = 1.8$$
$$BC = AD - AB - CD$$
$$BC = 8.1 - 1.8 - .9 = 5.4$$

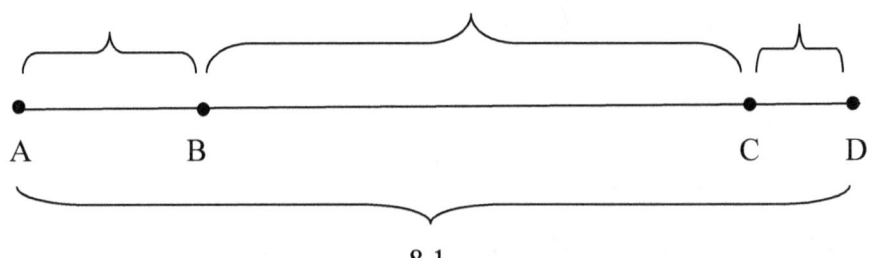

$$AC = AB + BC$$
$$AC = 1.8 + 5.4 = 7.2$$

$$\frac{BC}{AC} = \frac{5.4}{7.2} = \frac{3}{4}$$

Answer: (A) 3/4

15. If $f(x) = 5x - 2$ and $f(g(x)) = 10x$, find $g(x)$.

(A) $5x + 2$ (B) $2x + 2/5$ (C) $3x - 2/3$ (D) $x - 1/3$ (E) $3x + 2$

We are given that $f(x) = 5x - 2$. Plugging $g(x)$ into $f(x)$ yields

$$f(g(x)) = 5[g(x)] - 2$$

We are also given that $f(g(x)) = 10x$. This yields

$$5[g(x)] - 2 = 10x$$

$$5g(x) = 10x + 2$$

$$g(x) = 2x + 2/5$$

Answer: (B) $2x + 2/5$

16. In Mandy's Coffee Shop, for every a cups of coffee sold, $\sqrt{b}$ cups of tea are sold. Using the same ratio, how many cups of tea are sold when c cups of coffee are sold?

(A) $a\dfrac{\sqrt{b}}{c}$ (B) $\dfrac{\sqrt{b}}{ac}$ (C) $c\dfrac{\sqrt{b}}{a}$ (D) $\dfrac{ac}{\sqrt{b}}$ (E) $ac\sqrt{b}$

Forming the ratio of coffee to tea from the original information yields

$$\frac{\text{coffee}}{\text{tea}} = \frac{a}{\sqrt{b}}$$

Now, let t equal the unknown cups of tea and form the proportion:

$$\frac{c}{t} = \frac{a}{\sqrt{b}}$$

Cross-multiplying yields

$$c\sqrt{b} = at$$

$$\frac{c\sqrt{b}}{a} = t$$

Answer: (C) $\dfrac{c\sqrt{b}}{a}$

17. Find the equation of the line connecting points A(2, 3) and B(4, 9).

(A) $y = 2x - 4$ (B) $y = 2x + 3$ (C) $y = 2x - 4$ (D) $y = 3x - 3$ (E) $y = 3x + 1$

First, find the slope, m:

$$m = \frac{\Delta y}{\Delta x} = \frac{9-3}{4-2} = \frac{6}{2} = 3$$

The general form of a linear equation is $y = mx + b$, where m is the slope and b is the y-intercept.

Plugin point A, (2, 3):
$$y = 3x + b$$
$$3 = 3(2) + b$$
$$3 = 6 + b$$
$$b = -3$$

Go back to the general form of the linear equation.

$m = 3, b = -3$:
$$y = mx + b$$
$$y = 3x - 3$$

Answer: (D) $y = 3x - 3$

18. What is the approximate value of $\frac{2.7834}{7} \times 526$?

(A) 310 (B) 180 (C) 225 (D) 380 (E) 450

Round 2.7834 to 3 and 526 to 520. Note that we rounded 526 down to 520 because this balances the rounding up of 2.7834 to 3. Now, the expression becomes

$$\frac{3 \cdot 520}{7} \approx 223$$

Answer: (C) 225

19. If the temperature (T) of a star falls by 2% each million years and the temperature now is 1,000,000°, derive the formula for the temperature 10 million years from now.

(A) $1,000,000(.98)^{10}$ (B) $.02(1,000,000)^{10}$ (C) $1,000,000^9(.98)$
(D) $[(.02)(1,000,000)]^{10}$ (E) $1,000,000(.98)^9$

In a geometric sequence, $a_n = a_1 r^{n-1}$, where a_n is the last term, a_1 is the first term, r is the common ratio and n is the number of terms.

In 10 million years from now, 11 terms in the sequence will have passed. At the beginning of the first year, the temperature is 1,000,000°; and at the end of the first year, the temperature is 980,000°. So, in the first year, the temperature is measured twice (at the beginning and at the end of the year). In all other years, the temperature is measured only once, at the end of the year.

$$a_n = a_1 r^{n-1}$$

$a_1 = 1,000,000, r = .98, n = 11$:

$$a_n = (1,000,000)(.98)^{11-1} = (1,000,000)(.98)^{10}$$

Answer: (A) $1,000,000(.98)^{10}$

20. If the diameters of each of the smaller circles is 1/3 the diameter of the large circle, find the area in between the large circle and the smaller circles.

(A) $\dfrac{\pi r^2}{8}$ (B) $\dfrac{2\pi r^2}{9}$ (C) $\dfrac{\pi r^2}{27}$ (D) $\dfrac{5\pi r^2}{9}$ (E) $\dfrac{5\pi r^2}{18}$

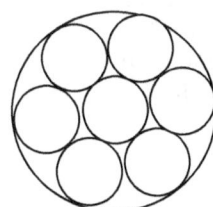

Let r = radius of the larger circle.
Then $r/3$ = radius of each of the smaller circles.
The area of the larger circle = πr^2.
The area of a smaller circle is $\pi\left(\dfrac{r}{3}\right)^2 = \dfrac{\pi r^2}{9}$.

There are 7 smaller circles, so the total area of the smaller circles is $\dfrac{7\pi r^2}{9}$.

The total area inside of the larger circle but outside the smaller circles is

$$\pi r^2 - \dfrac{7\pi r^2}{9} = \dfrac{2\pi r^2}{9}$$

Answer: (B) $\dfrac{2\pi r^2}{9}$

21. Find the equation of the perpendicular bisector of the line connecting the points A(–2, 0) and B(2, –8).

(A) $y = \frac{1}{2}x - 4$ (B) $y = 3x - 2$ (C) $y = \frac{1}{2}x + 3$ (D) $y = 2x + 4$ (E) $y = \frac{1}{2}x + 2$

Strategy: First, find the equation of the line, l_1, connecting the two points. Then find the midpoint of that line. The slope of a perpendicular to a line is the negative reciprocal of the slope of the original line.

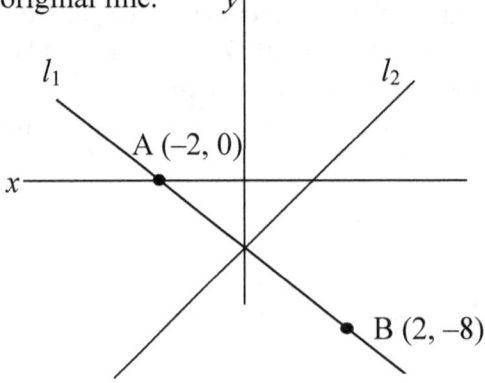

l_1:

(–2, 0):
1) $y = mx + b$
$0 = m(-2) + b$
$2m = b$
$m = \frac{b}{2}$

(2, –8):
2) $y = mx + b$
$-8 = (\frac{b}{2})2 + b$
$-8 = 2b$
$b = -4$
$m = \frac{b}{2}$
$= \frac{-4}{2}$
$= -2$
$y = -2x - 4$

The midpoint of the line connecting points A and B is

$$\left(\frac{-2+2}{2}, \frac{0+(-8)}{2}\right) = (0, -4)$$

The slope of the perpendicular bisector, l_2, is $-(\frac{1}{-2}) = \frac{1}{2}$.

The equation for line l_2, the perpendicular bisector of AB, where the midpoint is (0, –4) is

3) $y = mx + b$
$m = \frac{1}{2}$, (0, –4): $-4 = \frac{1}{2}(0) + b$
$b = -4$
$y = \frac{1}{2}x - 4$

Answer: (A) $y = \frac{1}{2}x - 4$

22. Multiply the two complex numbers $(4 + 2i)$ and $(3 - 5i)$.

(A) $6 - 3i$ (B) $14 + 12i$ (C) $16 + 2i$ (D) $22 - 14i$ (E) $16 - 5i$

$$(4 + 2i)(3 - 5i)$$

$$12 - 20i + 6i - 10i^2$$

$$12 - 14i - 10(-1)$$

$$12 - 14i + 10$$

$$22 - 14i$$

Answer: (D) $22 - 14i$

23. Solve for x in the equation

$$\frac{4}{\sqrt{2x+5}} = \frac{\sqrt{2x+5}}{x}$$

(A) 2 (B) $2\frac{1}{2}$ (C) $-3\frac{1}{2}$ (D) -3 (E) $-3\frac{1}{3}$

$$\frac{4}{\sqrt{2x+5}} = \frac{\sqrt{2x+5}}{x}$$

Cross-multiplying yields

$$4x = \sqrt{2x+5} \cdot \sqrt{2x+5}$$

$$4x = 2x + 5$$

$$2x = 5$$

$$x = \frac{5}{2} = 2\frac{1}{2}$$

Answer: (B) $2\frac{1}{2}$

24. An inspector checks b number of batteries every 10 minutes. If $.02c$ batteries fail every 1/2 hour, what fraction pass after 6 hours?

(A) $\dfrac{40b - .16c}{24b}$ (B) $\dfrac{2.6b - 10c}{36c}$ (C) $\dfrac{36b - .24c}{36b}$ (D) $\dfrac{16c - .04b}{24b - 2c}$ (E) $\dfrac{.3b - .24c}{.8b + c}$

6 hours = 6 × 60 minutes per hour = 360 minutes
6 hours = 6 × 2 = 12 half hours

$\dfrac{360 \text{ minutes}}{10 \text{ minutes}}$ = 36 ten-minute intervals in 6 hours

$36b$ total batteries checked in 6 hours

$12(.02c) = .24c$ total batteries that failed in 6 hours

total passing batteries = $\dfrac{\text{total batteries checked} - \text{total batteries failed}}{\text{total batteries checked}} = \dfrac{36b - .24c}{36b}$

$$\dfrac{36b - .24c}{36b}$$

Answer: (C) $\dfrac{36b - .24c}{36b}$

25. Find the distance between A(10, 11) and B (2, 5) on the xy-plane.

(A) 10   (B) 8   (C) $\sqrt{46}$ (D) $\sqrt{51}$ (E) 9

Note: The formula for the distance between two points (x_1, y_1) and (x_2, y_2) is

$$d = \sqrt{(x_2 - x_1)^2 - (y_2 - y)^2}$$

Applying this formula gives

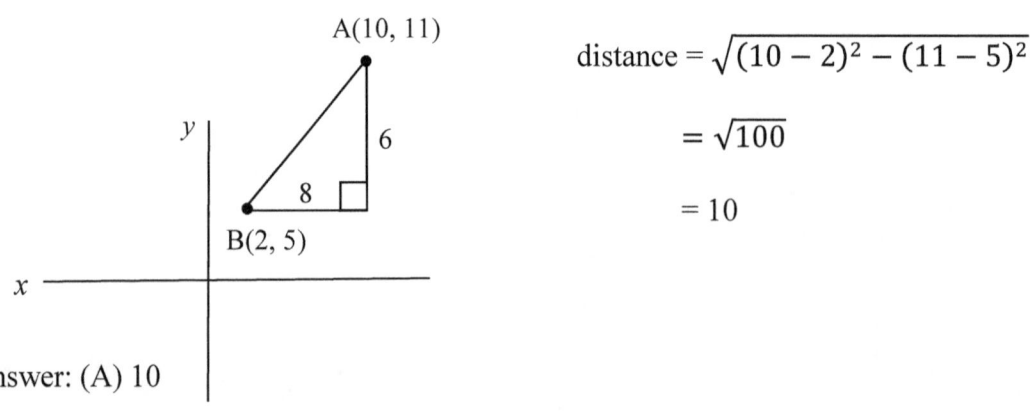

distance = $\sqrt{(10 - 2)^2 - (11 - 5)^2}$

     = $\sqrt{100}$

     = 10

Answer: (A) 10

Test 3

26. Given the statement "If $\sqrt{x+4} \geq 3$, then $x > 5$," which one of the following statements is a counterexample.

(A) $x < 5$ (B) $x = 2$ (C) $4 < x < 5$ (D) $x \leq 3$ (E) $x = 5$

A counterexample is an example that disproves a statement.

Let's look at (E) $x = 5$. We are given that when $\sqrt{x+4} \geq 3$, then $x > 5$. Plugging 5 into this inequality gives

$$\sqrt{5+4} \geq 3$$

$$\sqrt{9} \geq 3$$

$$3 = 3 \checkmark$$

In other words, x does not have to be greater than 5 ($x > 5$) for the original statement to be true. It could be 5.

Answer: (E), $x = 5$

27. Rationalize the denominator:

$$\frac{6-5i}{3+4i}$$

(A) $\frac{4+6i}{23}$ (B) $\frac{-2-39i}{25}$ (C) $\frac{5-6i}{14}$ (D) $\frac{2+3i}{26}$ (E) $\frac{7+8i}{18}$

Multiply the denominator $3 + 4i$ by its conjugate, $3 - 4i$. Also, multiply the numerator by the same conjugate so that the numerical value of the expression is not changed:

$$\frac{6-5i}{3+4i} \cdot \frac{3-4i}{3-4i}$$

$$\frac{18 - 24i - 15i + 20i^2}{9 - 12i + 12i - 16i^2}$$

$$\frac{18 - 24i - 15i + 20(-1)}{9 - 12i + 12i - 16(-1)}$$

$$\frac{-2 - 39i}{25}$$

Answer: (B) $\frac{-2-39i}{25}$

28. Find the vertical difference between the graphs of $y = 2x^2 - x + 4$ and $y = x^2 + 2x + 3$ at $x = -1$.

(A) 3 (B) 4 (C) 5 (D) 2 (E) 6

At $x = -1$: $y = 2x^2 - x + 4 = 2(-1)^2 - (-1) + 4 = 2 + 1 + 4 = 7$

At $x = -1$: $y = x^2 + 2x + 3 = (-1)^2 + 2(-1) + 3 = 1 - 2 + 3 = 2$

Now, the vertical difference is the difference between these two values:

$$7 - 2 = 5$$

Answer: (C) 5

29. If the 4th term of an arithmetic sequence is –9 and the first term is –3, find the median of the first 6 terms.

(A) –7 (B) –8 (C) –9 (D) –10 (E) 6

Let a_1 be the first term, –3.
Let a_n be the last term, –9.
Let n be the number of terms, 4.
Let d be the common difference.

Find the common difference. Then, find the first 6 terms and the median of all 6 terms.

$$a_n = a_1 + (n-1)d$$
$$-9 = -3 + (4-1)d$$
$$-9 = -3 + 3d$$
$$-6 = 3d$$
$$d = -2$$

So then nth term, $a_n = a_1 + (n-1)d$, becomes

$$a_n = -3 + (n-1)(-2) = -2n - 1$$

$$\text{median} = -8$$
$$-3, -5, -7, \downarrow -9, -11, -13$$

Answer: (B) –8

30. Determine the range of the function $y = 3 \cos 4x$.

(A) $-3 \leq y \leq 3$ (B) $-4 \leq y \leq 4$ (C) $-3 \leq y \leq 0$ (D) $-4 \leq y \leq 0$ (E) $-3 \leq y \leq 4$

$y = 3 \cos 4x$

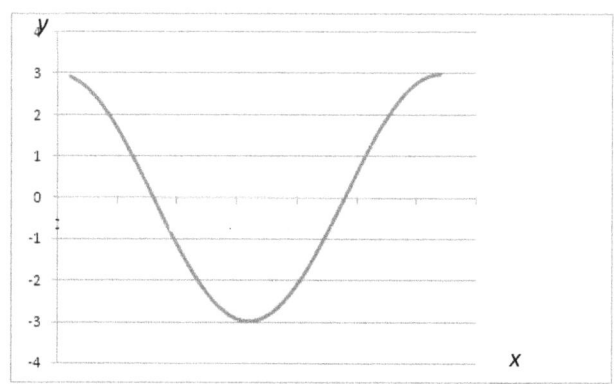

$f(x) = a \cos bx + c$, x is in radians

1. $|a|$ is the amplitude. So, $y = 3 \cos 4x$ has an amplitude of 3 ($= |3|$).

2. $\dfrac{2\pi}{b}$ ($b > 0$) is the period. So, the period is $2\pi/3$.

3. $+c$ indicates a move up. So, the vertical shift upwards is zero ($y = 3 \cos 4x + \mathbf{0}$).

4. $-c$ indicates a move down. So, the vertical shift downwards is zero ($y = 3 \cos 4x + \mathbf{0}$).

The range is $-3 \leq y \leq 3$.

Answer: (A) $-3 \leq y \leq 3$

ACT Math Personal Tutor

31. Miller's Department store sells a television set at a 20% discount. Jackson's, another department store, sells the same set for $25 more than Miller's discounted price. If Jackson's sells the set for $505, how much was the original non-discounted price?

(A) $560   (B) $580   (C) $600   (D) $620   (E) $490

Let x = original, non-discounted price.
Miller's price = $80\%x = .80x$
Jackson's price = $.80x + \$25$

Jackson's price is $505: $.80x + 25 = 505$
$.80x = 480$
$x = 600$

Answer: (C) $600

32. Given the following two equations, solve for s in terms of v.

(1) $r + 2s = t$
(2) $t - r = v$

(A) $s = \dfrac{v}{3}$ (B) $s = \dfrac{2v}{3}$ (C) $s = \dfrac{v}{4}$ (D) $s = \dfrac{v}{2}$ (E) $s = \dfrac{3v}{2}$

(1) $s = \dfrac{t-r}{2}$
(2) $r = t - v$

Substitute $t - v$ for r in (1).

(1) $s = \dfrac{t-(t-v)}{2}$
(1) $s = \dfrac{v}{2}$

Answer: (D) $s = \dfrac{v}{2}$

33. Reflect ΔABC with vertices A(3, 1), B(5, 1) and C(5, 3) along the line $y = x$ into its image A'B'C'. What are the co-ordinates of point B'?

(A) (1, 5)　　　(B) (2, 5)　　　(C) (4, 5)　　　(D) (1, 4)　　　(E) (2, 4)

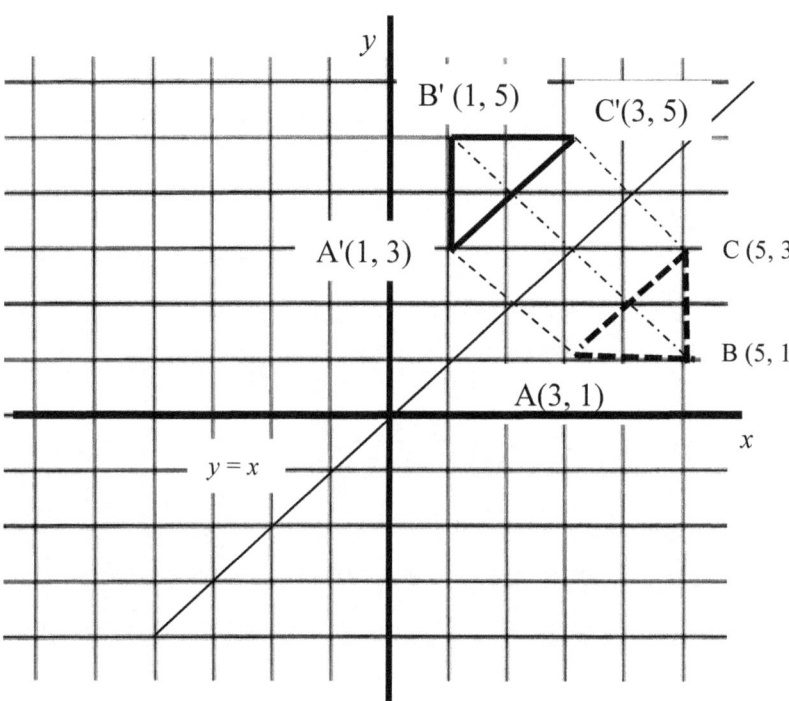

Draw lines from points A, B and C perpendicular to $y = x$. Measure the lengths of those lines and extend the lines the same length on the opposite side of the line $y = x$.

A simpler method is just to reverse the x and y co-ordinates.

Answer: (A) (1, 5)

34. A poll of people's reading preferences was taken and the results (expressed in degree measure) are pictured in the circle shown. If 1,800 people enjoy fiction, how many people in the poll enjoy history?

(A) 3,000   (B) 2,600   (C) 3,200   (D) 2,700   (E) 2,900

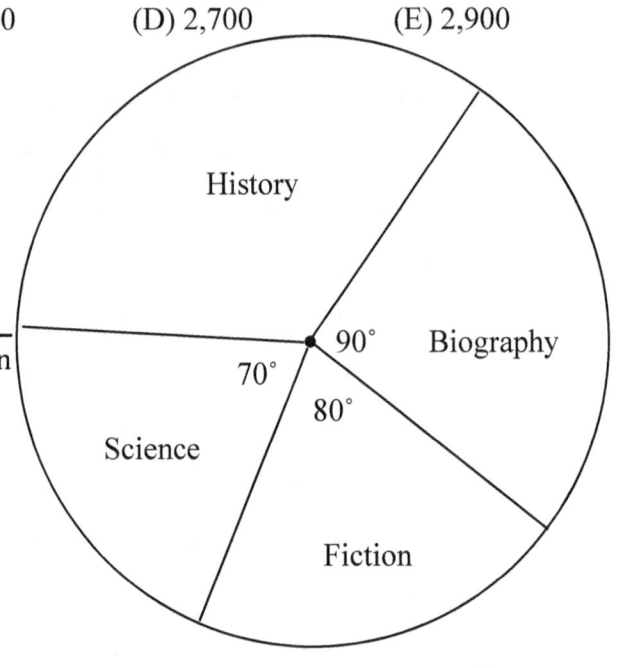

Number of degrees to represent history:
$360° - 90° - 80° - 70° = 120°$

Use proportions:

$$\frac{\text{degrees in history}}{\text{\# of persons in history}} = \frac{\text{degrees in fiction}}{\text{\# of persons in fiction}}$$

$$\frac{120°}{x} = \frac{80°}{1,800}$$

$80x = 216,000$

$x = 2,700$

Answer: (D) 2,700

35. Given the two equations $y + 2x = 0$ and $3^{xy} = \frac{1}{81}$, find the positive value of x.

(A) 2   (B) $\sqrt{2}$ (C) $\sqrt{3}$ (D) 3   (E) 1/3

(1) $y + 2x = 0$

(2) $3^{xy} = \frac{1}{81}$

(1) $y = -2x$

(2) $3^{x(-2x)} = 3^{-4}$

(2) $3^{-2x^2} = 3^{-4}$

(2) $-2x^2 = -4$

(2) $x^2 = 2$

(2) $x = \pm\sqrt{2}$

Answer: (B) $\sqrt{2}$

36. If $b = \sqrt[3]{c}$ and $8 \leq c \leq 125$ and $a = \dfrac{b}{c}$, find the greatest possible value for a.

(A) 5/8 (B) 1/2 (C) 7/8 (D) 1/4 (E) 3/4

$$a = \frac{b}{c} = \frac{\sqrt[3]{c}}{c} = \frac{1}{c^{2/3}}$$

To make the fraction $\dfrac{1}{c^{2/3}}$ as large as possible, make the denominator as small as possible (for example $\dfrac{1}{2} > \dfrac{1}{3}$). Since $8 \leq c \leq 125$, choose $c = 8$:

$$a = \frac{1}{c^{2/3}} = \frac{1}{8^{2/3}} = \frac{1}{2^2} = \frac{1}{4}$$

Answer: (D) 1/4

37. If the volume of a cube is 125 cubic inches, what is its total surface area?

(A) 120 sq in (B) 225 sq in (C) 150 sq in (D) 180 sq in (E) 210 sq in

$V = s^3 = 125$ cu in
$s = \sqrt[3]{125} = 5$ in

1 side = 5 × 5 = 25 sq in
(6 sides) × 25 = 150 sq in

Answer: (C) 150 sq in

38. The table below indicates values of the function $f(x)$ for given values of x.

x	1	2	3	4
$f(x)$	2	5	10	17

Which function is most likely represented by this table?

(A) $3x^2 - 1$ (B) $x^2 + 2$ (C) $2x^2 + 3$ (D) $x^2 - x + 1$ (E) $x^2 + 1$

(E): $f(1) = 1^2 + 1 = 2$
$f(2) = 2^2 + 1 = 5$
$f(3) = 3^2 + 1 = 10$
$f(4) = 4^2 + 1 = 17$

Answer: (E) $x^2 + 1$

39. Jack goes to the Nuart Theater every 2nd day at noon, Myra goes to the same theater every 3rd day at noon, and Jose goes to the same theater every 4th day at noon. If they all go to the Nuart on the same day, when is the earliest day they will they all meet again at the theater?

(A) 8 days   (B) 13 days   (C) 10 days   (D) 11 days   (E) 9 days

Jack: 1, 3, 5, 7, 9, 11, **13**,...
Myra: 1, 4, 7, 10, **13**, 16,...
Jose: 1, 5, 9, **13**, 17,...

The least common multiple is 13.

Answer: (B) 13 days

40. If the average 18-year old would reduce her daily intake of sugar by 1/5, she would still consume 10 more grams of sugar a day than the average 26-year old. If the average 26-year old consumes 70 grams of sugar per day, how many grams of sugar does the average 18-year old consume?

(A) 300   (B) 900   (C) 400   (D) 100   (E) 80

Let x = the daily grams of sugar consumed by the average 18-year old.

The average 26-year old consumes 70 grams of sugar per day.

If the average 18-year old would reduce her sugar by 1/5, she would consume 10 grams more sugar than the average 26-year old.

$$x - \frac{1}{5}x = 70 + 10$$

$$\frac{4}{5}x = 80$$

Multiply by 5:

$$4x = 400$$

Divide by 4:

$$x = 100$$

Answer: (D) 100

Test 3

41. The electrical resistance of a wire, r, varies directly as its length, l, and inversely as the square of its diameter, d. The resistance of a wire 400 inches in length with a diameter of 1/4 of an inch is 32 ohms. Find the resistance of a wire 200 inches in length with a diameter of 1/8 of an inch.

(A) 20 ohms   (B) 64 ohms   (C) 44 ohms   (D) 32 ohms   (E) 28 ohms

By definition *r varies directly as l* means $r = k \cdot l$, for some constant k. Further, by definition *r varies inversely as square of d* means $r = k\frac{1}{d^2}$, for some constant k. Combining these two definitions yields

$$r = k\frac{l}{d^2}$$

First, plugin the given information to find the constant, k.

$r = 32, l = 400, d = 1/4$:

$$32 = k\frac{400}{\left(1/4\right)^2}$$

$$32 = k\frac{400}{1/16}$$

$$32 = k \cdot 400 \cdot \frac{16}{1}$$

$$32 = k \cdot 6400$$

$$k = \frac{1}{200} = .005$$

So, the formula becomes

$$r = .005\frac{l}{d^2}$$

Now, plugin the given information to find the r.

$k = .005, l = 200: d = 1/8$:

$$r = .005\frac{200}{\left(1/8\right)^2}$$

$$r = \frac{1}{1/64}$$

$r = 64$

Answer: (B) 64 ohms

42. If $|x| y^{2/3} = 80$ and $x = -5$, what is the value of y?

(A) 48 (B) 32 (C) 16 (D) 64 (E) 80

$$|x| y^{2/3} = 80$$
$$|-5| y^{2/3} = 80$$
$$5 y^{2/3} = 80$$
$$y^{2/3} = 16$$
$$(y^{2/3})^{3/2} = (16)^{3/2}$$
$$y = 64$$

Answer: (D) 64

43. The length of a rectangle is 4 inches more than its width. If the length is doubled and the width is reduced by 2 inches, the area of the new rectangle is 20 square inches greater than the area of the original rectangle. Find the length of the original rectangle.

(A) 10 (B) 12 (C) 9 (D) 8 (E) 14

original rectangle — sides x and $x + 4$

new rectangle — sides $x - 2$ and $2(x + 4)$

$$(\text{new rectangle}) = (\text{old rectangle}) + 20$$
$$(x - 2)(2x + 8) = x(x + 4) + 20$$
$$2x^2 + 8x - 4x - 16 = x^2 + 4x + 20$$
$$2x^2 + 4x - 16 = x^2 + 4x + 20$$
$$x^2 - 36 = 0$$
$$(x + 6)(x - 6) = 0$$
$$x + 6 = 0 \text{ or } x - 6 = 0$$
$$x = -6 \text{ or } x = 6$$

Reject -6 because x is a distance, which cannot be negative.

$$\text{Original length} = x + 4 = 6 + 4 = 10$$

Answer: (A) 10

44. Find the 48th digit of the repeating decimal 0.26734....

(A) 2 (B) 6 (C) 7 (D) 3 (E) 4

The numbers (26734) repeat in blocks of 5. So, divide 48 by 5.

If the remainder is 1, the digit is 2.
If the remainder is 2, the digit is 6.
If the remainder is 3, the digit is 7.
If the remainder is 4, the digit is 3.
If the remainder is 0, the digit is 4.

Now, 5 divides into 48 nine times with a remainder of 3.

So, the 48th digit is 7.

Answer: (C) 7

45. An equilateral triangle is inscribed in a circle of radius 4 as shown. Find the shaded area. Leave your answer in terms of π and radicals.

(A) $\dfrac{3\pi + 4\sqrt{3}}{4}$ (B) $\dfrac{16\pi - 12\sqrt{3}}{3}$ (C) $\dfrac{12\pi + 2\sqrt{2}}{3}$ (D) $\dfrac{12\pi - 2\sqrt{3}}{4}$ (E) $\dfrac{16\pi - 4\sqrt{3}}{4}$

Area $\Delta = \dfrac{1}{2} ab \sin C$, where a and b are sides of the triangle and C is the included angle.

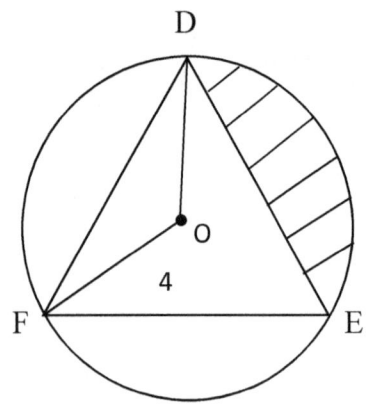

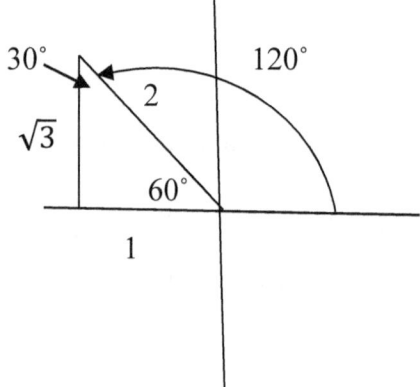

$m\angle DOF = \dfrac{1}{3} \times 360° = 120°$

$\sin 120° = \sin 60° = \dfrac{\sqrt{3}}{2}$

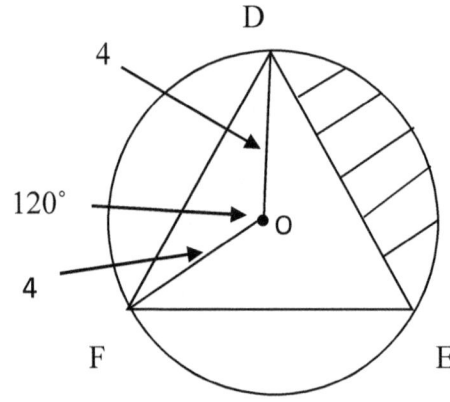

Area $\Delta DOF = \dfrac{1}{2} ab \sin \angle DOF$
$= \dfrac{1}{2}(4)(4) \sin 120°$
$= 8 \sin 60°$
$= 8 \times \dfrac{\sqrt{3}}{2} = 4\sqrt{3}$

Area $\Delta DEF = 3 \times$ Area $\Delta DOF = 3 \times 4\sqrt{3}$
$= 12\sqrt{3}$

Area Circle $O = \pi r^2$
$= \pi 4^2 = 16\pi$

Shaded Area $= \dfrac{\text{Area of Circle} - \text{Area of Triangle DEF}}{3} = \dfrac{16\pi - 12\sqrt{3}}{3}$

Answer: (B) $\dfrac{16\pi - 12\sqrt{3}}{3}$

46. Find the value of $x + y$ in the matrix addition problem shown.

$$\begin{bmatrix} 2 & 4 \\ 3 & y \end{bmatrix} + \begin{bmatrix} 5 & x \\ 2 & 7 \end{bmatrix} = \begin{bmatrix} 7 & 10 \\ 5 & 6 \end{bmatrix}$$

(A) −1   (B) 4   (C) 5   (D) 3   (E) −2

Adding the two matrices on the left yields

$$\begin{bmatrix} 2+5 & 4+x \\ 3+2 & y+7 \end{bmatrix} = \begin{bmatrix} 7 & 10 \\ 5 & 6 \end{bmatrix}$$

$$\begin{bmatrix} 7 & 4+x \\ 5 & y+7 \end{bmatrix} = \begin{bmatrix} 7 & 10 \\ 5 & 6 \end{bmatrix}$$

Two matrices are equal if and only if corresponding entries are equal. This yields the following system of two equations:

(1) $4 + x = 10$
$x = 6$
(2) $y + 7 = 6$
$y = -1$

$x + y = 6 + (-1) = 5$

Answer: (C) 5

47. A ball is dropped from a height of 24 feet. If it continues to bounce up 1/2 the dropped height, how many times does it hit the ground when the height after the last bounce measures 3/4 of a foot?

(A) 1 bounce   (B) 2 bounces   (C) 3 bounces   (D) 4 bounces   (E) 5 bounces

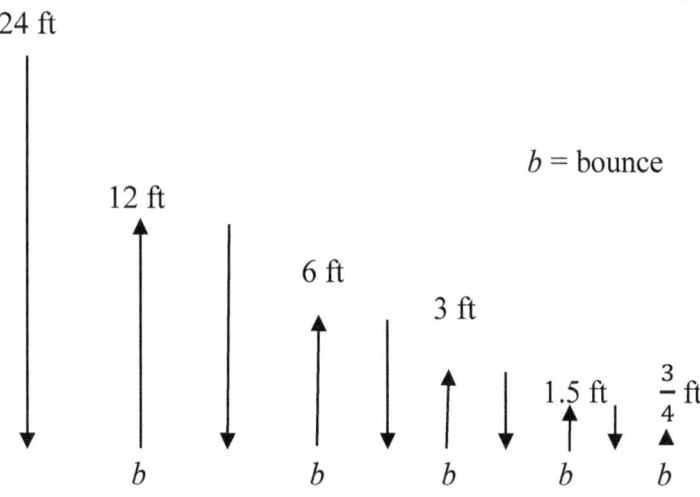

Each letter b in the figure signifies a bounce. There are 5 b's in the figure. Hence, there are 5 bounces.

Answer: (E) 5 bounces

ACT Math Personal Tutor

48. Which of the choices has the same roots as $f(x) = 2x^2 - 2x - 12$?

(A) $x^2 - x - 6$ (B) $x^2 + 6x - 2$ (C) $3x^2 - 2x - 4$ (D) $x^2 + 4x - 3$ (E) $2x^2 + 3x - 5$

$$2x^2 - 2x - 12 = 0$$
$$2(x^2 - x - 6) = 0$$
$$2(x - 3)(x + 2) = 0$$
$$x - 3 = 0 \text{ or } x + 2 = 0$$
$$x = 3 \text{ or } x = -2$$

Now, look at Choice (A):

$$x^2 - x - 6 = 0$$
$$(x - 3)(x + 2) = 0$$
$$x = 3 \text{ or } x = -2$$

Answer: (A) $x^2 - x - 6$

49. There are 4 positions available for the chess team. If 9 students apply, in how many ways can the positions be filled?

(A) 2,016   (B) 4,268   (C) 3,024   (D) 4,561   (E) 2,884

There are four positions available. There are 9 possible candidates for the first position. Once that position is filled, there are 8 possible candidates for the second position. And so on....

$$\underline{9} \times \underline{8} \times \underline{7} \times \underline{6} = 3{,}024$$

Answer: (C) 3,024

50. We want to get 100 ounces of a 44% solution of alcohol by mixing an 80% solution of alcohol with a 20% solution of alcohol. How many ounces of the 20% solution should we include?

(A) 20 ounces (B) 50 ounces (C) 30 ounces (D) 40 ounces (E) 60 ounces

Most mixture problem are solved with the following equation:

(Amount of stuff in one solution) + (Amount of stuff in another solution) = (Amount of stuff in the combined mixture)

For this problem, the stuff is alcohol. And we find the amount of alcohol in a solution by multiplying the amount of the solution by the percentage of alcohol in that solution.

Let x = the number of ounces of the 80% solution.
Let $100 - x$ = the number of ounces of the 20% solution.

(Alcohol in 80% solution) + (Alcohol in 20% solution) = (Alcohol in 44% solution)

$$.80x + .20(100 - x) = .44(100)$$

Multiply by 100:

$$80x + 20(100 - x) = 44(100)$$
$$80x + 2000 - 20x = 4400$$
$$60x = 2400$$
$$x = 40$$
$$100 - x = 60$$

Answer: (E) 60 ounces

51. Simplify

$$\frac{2x + 8}{2x^2 + 6x - 8}$$

(A) $\dfrac{2x}{x - 4}$ (B) $\dfrac{1}{x - 1}$ (C) $\dfrac{3x}{2x + 1}$ (D) $\dfrac{2}{3x - 2}$ (E) $\dfrac{4x}{2x + 1}$

$$\frac{2x + 8}{2x^2 + 6x - 8}$$
$$\frac{2(x + 4)}{2(x^2 + 3x - 4)}$$
$$\frac{x + 4}{(x + 4)(x - 1)}$$
$$\frac{1}{x - 1}$$

Answer: (B) $\dfrac{1}{x-1}$

52. Determine the axis of symmetry of the graph represented by the function $f(x) = (2x + 1)(x + 3)$.

(A) –3/4   (B) –7/4   (C) 5/8   (D) 9/7   (E) –8/5

$f(x) = (2x + 1)(x + 3) = 2x^2 + 6x + x + 3$
$= 2x^2 + 7x + 3$

The x-value of the axis of symmetry is $-b/2a = -7/2(2) = -7/4$.

Answer: (B) –7/4

53. A particular type of plant has been dying at the rate of 3% per year. If there are 332,520 plants at the current time, how many plants were there 6 years ago? Use the formula $A = i(1 – r)^n$, where A = the final number, i = the initial population, r = % decrease per time period and n = the number of time periods. Round off to the nearest whole number

(A) 375,420   (B) 401,426   (C) 399,197   (D) 385,176   (E) 380,526

$A = 332{,}520, r = .03, n = 6$

$$A = i(1 – r)^n$$
$$332{,}520 = i(1 – .03)^6$$
$$332{,}520 = i(.97)^6$$
$$i = 399{,}197$$

Answer: (C) 399,197

54. The average of three whole numbers, a, b and c is 24. If b is 2 more than a and c is 14 more than b, what is the value of a?

(A) 6   (B) 8   (C) 18   (D) 22   (E) 16

a
$b = a + 2$
$c = b + 14 = (a + 2) + 14 = a + 16$

Forming the average of a, b, c gives

$$\frac{a + b + c}{3} = \frac{a + (a + 2) + (a + 16)}{3} = 24$$
$$\frac{3a + 18}{3} = 24$$
$$3a + 18 = 72$$
$$3a = 54$$
$$a = 18$$

Answer: (C) 18

Test 3

55. If $f(x) = 2x^2 + 3$ and $g(x) = x - 4$, find the value of $g\left(f(\sqrt{2})\right)$.

(A) 3   (B) 4   (C) 5   (D) 6   (E) 7

$f(x) = 2x^2 + 3$
$f(\sqrt{2}) = 2(\sqrt{2})^2 + 3 = 2(2) + 3 = 4 + 3 = 7$

$g(x) = x - 4$
$g(7) = 7 - 4 = 3$

Answer: (A) 3

Method II:

Most mathematical expressions are read from left to right, just as we read a sentence. But the composition of functions is read from inner parentheses out:

$$g\left(f(\sqrt{2})\right) = g\left(2(\sqrt{2})^2 + 3\right) = g(7) = 7 - 4 = 3$$

56. Identify one point where the graph of the line $f(x) = 0$ cuts through the graph of $g(x) = x^3 - 2x$?

(A) (2, 0)   (B) (0, $\sqrt{2}$)   (C) (0, 2)   (D) ($\sqrt{2}$, 0)   (E) ($\sqrt{3}$, 0)

When the functions touch, they have the same y-value:

$$f(x) = g(x)$$
$$0 = x^3 - 2x$$
$$0 = x(x^2 - 2)$$
$$x = 0 \text{ or } x^2 - 2 = 0$$
$$x = 0 \text{ or } x^2 = 2$$
$$x = 0 \text{ or } x = \pm\sqrt{2}$$

The line $y = 0$ cuts through the graph of $x^3 - 2x$ at the points (0, 0), (+$\sqrt{2}$, 0), (–$\sqrt{2}$, 0).

Answer: (D) ($\sqrt{2}$, 0)

57. In a survey of the ice cream preferences of 80 persons, 18 liked vanilla alone, 11 liked chocolate alone, and 15 liked a mix of chocolate and strawberry. Six liked strawberry alone, and the rest liked a mix of strawberry and vanilla. How many liked a mix of vanilla and strawberry?

(A) 31 (B) 36 (C) 19 (D) 23 (E) 30

Let's drawn a Venn Diagram:

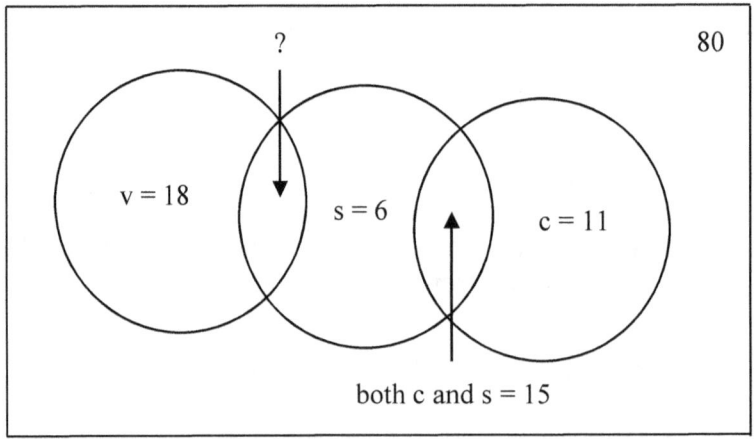

a mix of vanilla and strawberry = 80 – 18 – 6 – 15 – 11 = 30

Answer: (E) 30

Test 3

58. Which of the following graphs represents the equation $y = \dfrac{1}{x^2}, x > 0$?

(A)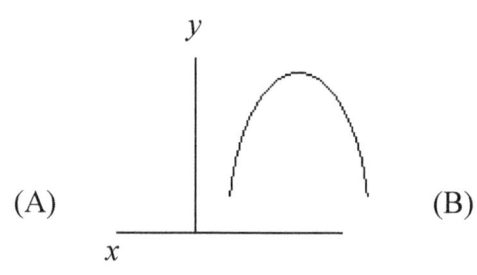
(B)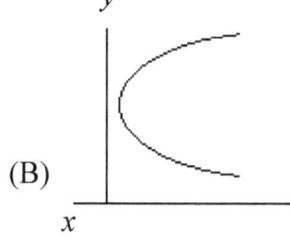
(C)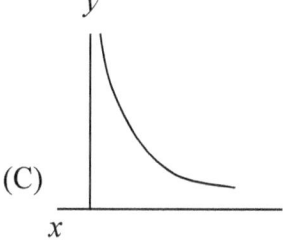

(D)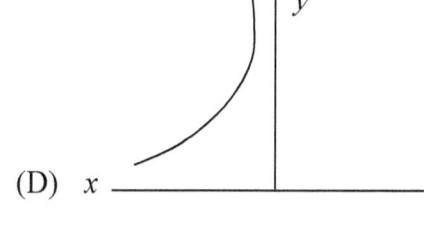
(E)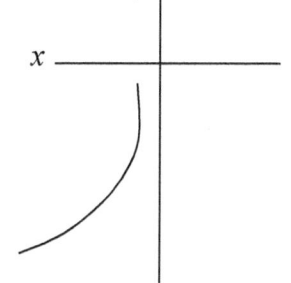

The easier way is to use your graphing calculator.

Another method is to create a simple table and plot the results.

$$y = \dfrac{1}{x^2}, x > 0$$

x	y
1	1
2	$\frac{1}{4}$
3	$\frac{1}{9}$

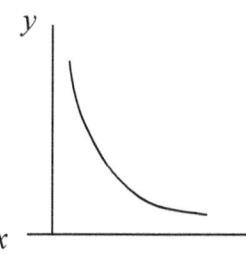

Answer: (C)

59. Find the value of $a - b$ in the matrix subtraction problem shown.

(A) 3   (B) –4   (C) 6   (D) –1   (E) –2

$$\begin{bmatrix} 2 & 4 \\ 5 & -1 \end{bmatrix} - \begin{bmatrix} a & 3 \\ 6 & 2 \end{bmatrix} = \begin{bmatrix} 6 & 7 \\ -1 & b \end{bmatrix}$$

Subtracting the two matrices on the left yields

$$\begin{bmatrix} 2-a & 4-3 \\ 5-6 & -1-2 \end{bmatrix} = \begin{bmatrix} 6 & 7 \\ -1 & b \end{bmatrix}$$

$$\begin{bmatrix} 2-a & 1 \\ -1 & -3 \end{bmatrix} = \begin{bmatrix} 6 & 7 \\ -1 & b \end{bmatrix}$$

Two matrices are equal if and only if corresponding entries are equal. This yields the following system of two equations:

(1) $2 - a = 6$
$a = -4$

(2) $b = -3$

$a - b = -4 - (-3)$
$= -1$

Answer: (D) –1

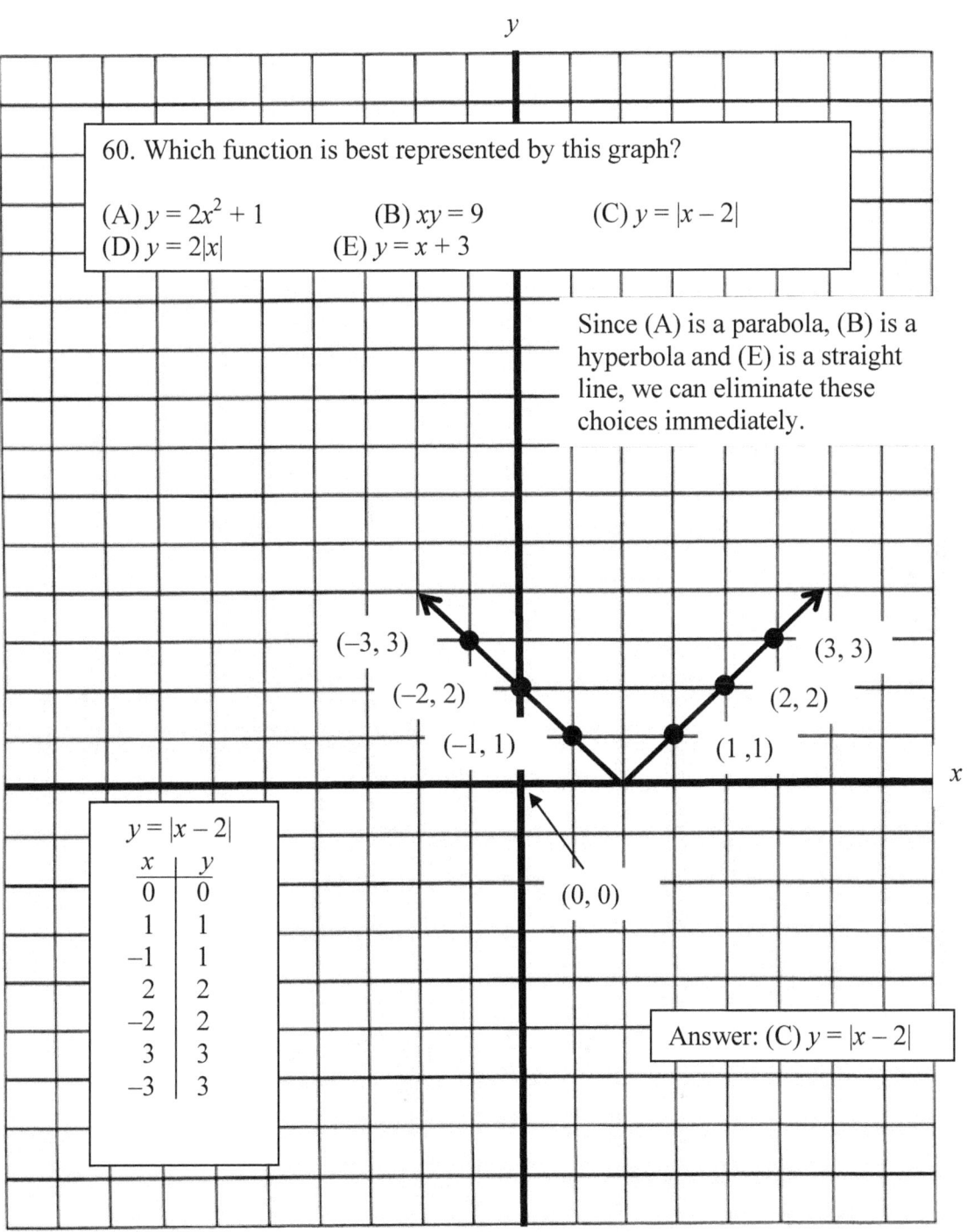

60. Which function is best represented by this graph?

(A) $y = 2x^2 + 1$ (B) $xy = 9$ (C) $y = |x - 2|$
(D) $y = 2|x|$ (E) $y = x + 3$

Since (A) is a parabola, (B) is a hyperbola and (E) is a straight line, we can eliminate these choices immediately.

$y = |x - 2|$

x	y
0	0
1	1
−1	1
2	2
−2	2
3	3
−3	3

Answer: (C) $y = |x - 2|$

Answer Key to Test 3

1. A
2. D
3. B
4. B
5. E
6. A
7. C
8. A
9. B
10. C
11. C
12. B
13. E
14. A
15. B
16. C
17. D
18. C
19. A
20. B
21. A
22. D
23. B
24. C
25. A
26. E
27. B
28. C
29. B
30. A
31. C
32. D
33. A
34. D
35. B
36. D
37. C
38. E
39. B
40. D
41. B
42. D
43. A
44. C
45. B
46. C
47. E
48. A
49. C
50. E
51. B
52. B
53. C
54. C
55. A
56. D
57. E
58. C
59. D
60. C

Test 4

1. Find the value of a in the equation $16^{a/2} = 8$.

   (A) 2/3   (B) 1/3   (C) 3/4   (D) 3/2   (E) –2/3

2. Let U = {a, b, c, d, e, f, g}, △ = {a, b, c, d}, ◯ = {c, d, e, f}, and ▢ = {a}. Find ~{▢ ∩ ◯} ∩ {◯ ∩ ~△}.

   (A) {b, c, d}   (B) {d, e, f, g}   (C) {e, f}   (D) {a, b, c, d}   (E) {d, e, f, g}

3. Subtract $5 - \sqrt{-32}$ from $3 + \sqrt{-18}$.

   (A) $3 + 2i$ (B) $-2 + 7i\sqrt{2}$ (C) $5 - 4\sqrt{-2}$ (D) $3 - 4\sqrt{-3}$ (E) $3 + 5i\sqrt{3}$

4. Lucy purchases 100 shares of stock of ABC Corporation at $7 per share. The stock pays an annual dividend of 2.50% on the original purchase price of $7. The first year each share increased in value by 3%. The second year each share dropped 2% and the third year each share increased by 6%. Including dividends, how much is Lucy's investment worth at the end of 3 years? Select the closest answer.

   (A) $909   (B) $854   (C) $746   (D) $917   (E) $802

5. C and D are different positive integers. Find the value of C in this addition problem.

   ```
       D C
   +   C D
     1 C 2
   ```

   (A) 2   (B) 1   (C) 7   (D) 3   (E) 8

6. Which one of the following points is on the graph of $f(x) = x^2 - x - 6 = 0$?

   (A) (0, 1)   (B) (1, –6)   (C) (3, 0)   (D) (3, 4)   (E) (4, 5)

7. Which equation is possibly represented by this graph?

(A) $y = x^2 + 2$ (B) $x - 2 = y$ (C) $y^2 = x + 1$ (D) $y = \pm\sqrt{x} + 2$ (E) $y = -x^2 + 2$

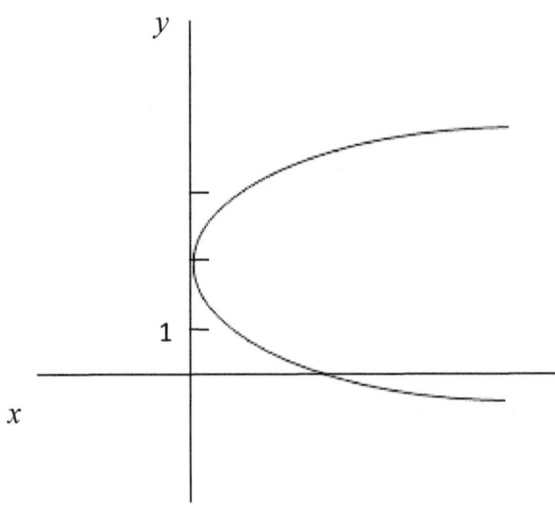

8. A harmonic sequence is a sequence of numbers whose reciprocals form an arithmetic sequence. If there are 5 terms in a harmonic sequence whose first term is 1/5 and whose last term is 1/21, what is the third term in the harmonic sequence?

(A) 1/13   (B) 3/11   (C) 1/7   (D) 2/7   (E) 2/11

9. Joel saved $20 for his Father's Day gift. His sister, Marilyn, saved 40% of the total amount of the gift. If they still needed $10, what was the total amount of the gift?

(A) $30   (B) $60   (C) $40   (D) $20   (E) $50

10. Mr. Jenkins invests $9,000 at a 5% annual interest rate without compounding for 2 years. He then takes the total and invests it for 3 additional years. If he has $11,088 at the end of the 3 years and the interest rate was not compounded, what was the interest rate for the second investment?

(A) 3%   (B) 6%   (C) 4%   (D) 3.5%   (E) 4.5%

11. Multiply:

$$\frac{x^2 - x - 6}{2x - 6} \cdot \frac{4x - 12}{x^2 + 5x + 6}$$

(A) $\dfrac{2(x+2)}{x-3}$ (B) $\dfrac{3(x+3)}{x+2}$ (C) $\dfrac{x-3}{x+3}$ (D) $\dfrac{2(x-3)}{x+3}$ (E) $\dfrac{3(2x+1)}{x-2}$

12. Below is a list of books borrowed in a month from the Gotham Library. The percents of the total number of books borrowed are listed, but only one category, biography, lists the actual number of books borrowed. How many history books were borrowed?

(A) 5,800 (B) 4,500 (C) 6,000 (D) 5,000 (E) 5,400

Books Borrowed from Gotham Library		
Category	**% of Total**	**Number**
mysteries	40	
poetry	5	
biography	20	4,000
history	25	
science	10	

13. Given the two equations $y = 2x - 5$ and $x = \dfrac{y}{3} + \dfrac{7}{3}$, where do their graphs intersect?

(A) (2, 1) (B) (3, –1) (C) (2, –1) (D) (2, –3) (E) (1, 4)

14. If $a = 6r^2$ and the ratio of $a : b$ is $2 : 7$, what is the value of b in terms of r?

(A) $14r$ (B) $14r^2$ (C) $18r$ (D) $16r^2$ (E) $21r^2$

15. There are a residents of Middle City above age 30 and b residents age 30 or under. If 600 children are born and 400 people above 30 leave the city, what fraction of the total new population represents the number of new residents above age 30?

(A) $\dfrac{a + 600}{a + b + 400}$ (B) $\dfrac{a - 400}{a - 600 + b}$ (C) $\dfrac{a + b - 600}{b + 400}$ (D) $\dfrac{a - 400}{a + b + 200}$

(E) $\dfrac{2a + b}{300 - a}$

16. If a possible span of the area, A, of a semi-circle is $12.5\pi \leq A \leq 18\pi$, what is a possible span of values for the radius, r?

(A) $5 \leq r \leq 6$ (B) $3 < r < 4$ (C) $4\ r < 5$ (D) $2 \leq r \leq 3$ (E) $5 < r < 7$

17. The units' digit of a three-digit number is 8. The hundreds' digit is 3. Find the tens' digit if a multiple of the number is 1014.

(A) 2   (B) 3   (C) 4   (D) 5   (E) 6

18. Two points, (4, 5) and (2, 7), lie on the circumference of a circle at the ends of a diameter. Find the center of the circle.

(A) (2, 4)   (B) (3, 5)   (C) (4, 6)   (D) (3, 4)   (E) (3, 6)

19. In the figure, triangle ABC is isosceles and AB ≅ BC. DE || FG || AC. $m \angle BHG = 115°$. Find the measure of $\angle B$.

(A) 50°   (B) 30°   (C) 40°   (D) 60°   (E) 70°

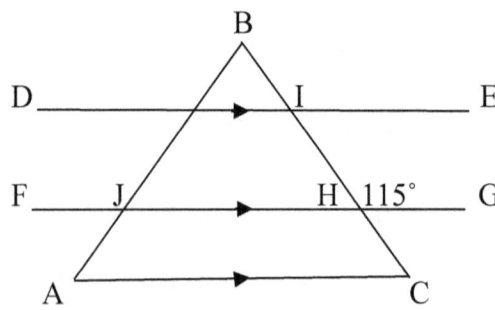

20. Given horizontal line l, how many points are located 4 inches away from that line and 5 inches away from point a on that line?

(A) 1   (B) 3   (C) 4   (D) 2   (E) none

21. Diamond Airlines flies twice as many passengers per day as Euclid Air. Euclid flies 1/3 as many passengers per day as Maloney Air. If the three airlines together fly 30,000 passengers per day, how many passengers does Maloney Air fly per day?

(A) 20,000   (B) 15,000   (C) 10,000   (D) 12,000   (E) 18,000

22. If $90° < \angle B < 180°$ and sin B = 1/2, find sec B.

(A) $\dfrac{\sqrt{3}}{2}$ (B) $\dfrac{1}{\sqrt{3}}$ (C) 2 (D) $\dfrac{-2}{\sqrt{3}}$ (E) $\dfrac{-\sqrt{3}}{2}$

23. Find the area of the triangle in the diagram shown.

(A) $de + ef$ (B) $de + e$ (C) $fe + \dfrac{1}{2}d$ (D) $df + e$ (E) $de + f$

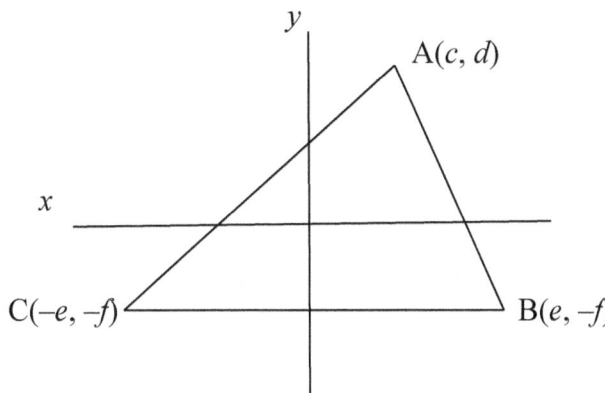

24. If x = arctangent (-2.5), where is x located?

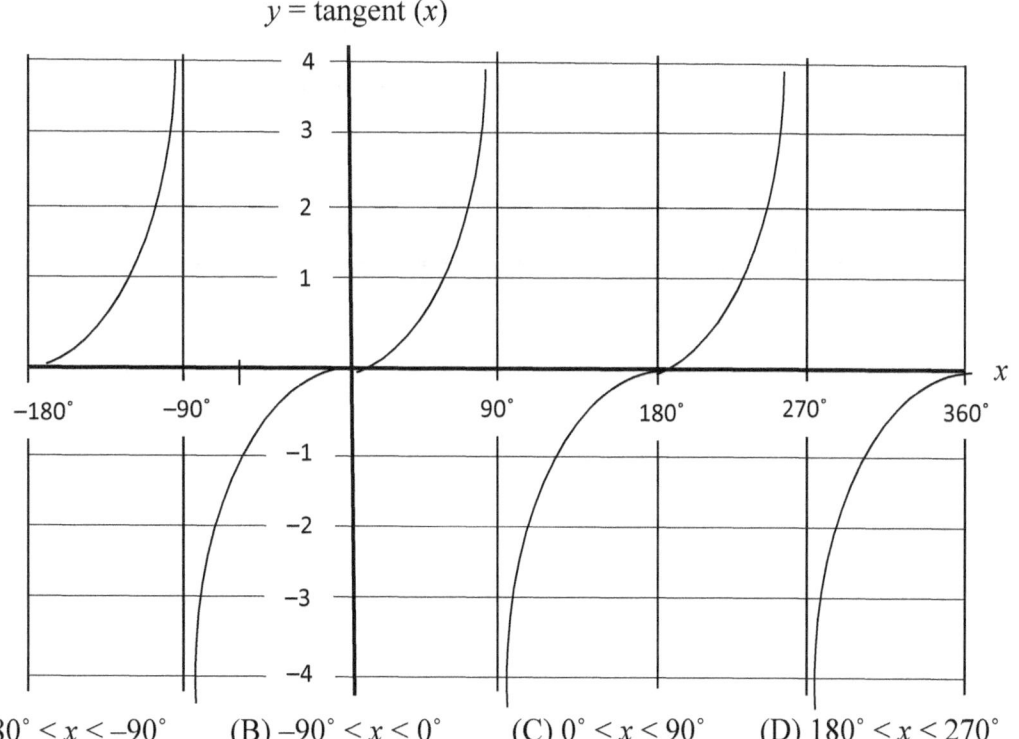

A) $-180° < x < -90°$ (B) $-90° < x < 0°$ (C) $0° < x < 90°$ (D) $180° < x < 270°$
(E) $180° \leq x \leq 270°$

25. The XYZ Corporation manufactures auto tires. The manufacturing cost of each tire, C, varies jointly as the cost of rubber per tire, R, and the cost of labor per tire, L. The cost of manufacturing a tire is $18 when the labor is $4 and rubber $3. Find the cost of manufacturing one tire if labor goes up to $5 and rubber drops to $2.

(A) $15 (B) $16 (C) $17 (D) $18 (E) $19

26. Jenny has to register for 3 college courses. She can take 1 out of 6 math classes, 1 out of 4 history classes, and 2 out of 5 science classes. How many different combinations of 3 courses can she choose?

(A) 480 (B) 120 (C) 240 (D) 180 (E) 600

27. What is the maximum point of the graph of the function $f(x) = -x^2 - 6x + 5$?

(A) (3, 10) (B) (-3, 14) (C) (6, 10) (D) (7, 12) (E) (2, 12)

28. The letters a, b and c represent positive integers such that $2c = 3a$ and $-4b = c$. Find the value of a when $a + b + c = 3$.

(A) $2\frac{3}{11}$ (B) $3\frac{4}{9}$ (C) $3\frac{8}{11}$ (D) $3\frac{5}{7}$ (E) $1\frac{7}{17}$

29. In an analog clock, which is functioning exactly, how many degrees are there between the hour hand and the minute hand at 9:26?

(A) 116° (B) 138° (C) 127° (D) 135° (E) 142°

30. Select the graph which represents the answer for the inequality $x^2 - 2x - 8 \geq 0$.

(A)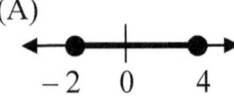
(B)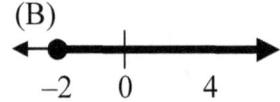
(C)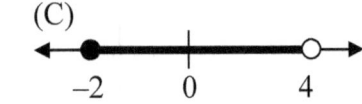

(D)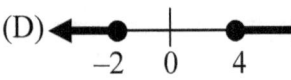
(E)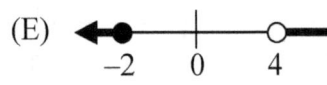

31. Jocelyn makes a deposit of $1,000 in a bank account. If the bank pays a rate of 2% a year and the deposit is compounded four times a year, how much will she have in her account at the end of six years? Use the formula

$$A = P\left(1 + \frac{r}{n}\right)^{nt}$$

where A = the final amount in the account, P = the initial principle, r = the annual rate, t = the number of years involved and n = the number of times a year compounding takes place.

(A) $1,000(1.05)^4$ (B) $1,000(1.05)^{24}$ (C) $1,000(1.005)^{10}$
(D) $1,000(1.005)^{24}$ (E) $1,000(24)^{.05}$

32. In the function shown, if $f(k) = 2$, then which one of the following could be a value of k?

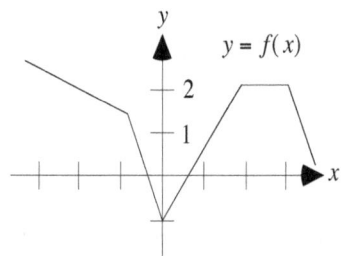

(A) –1 (B) 0 (C) 0.5 (D) 2.5 (E) 4

33. Simplify

$$\frac{\csc \theta}{\sin \theta} \cdot (1 - \cos 2\theta)$$

(A) 1 (B) 2 (C) sin θ (D) 3 (E) $\csc^2 \theta$

34. The letters *a*, *b*, *c* and *d* represent integers in ascending order of size. The letters *a* and *b* represent negative integers, while *c* and *d* represent positive integers. Which choice is the smallest?

(A) $\dfrac{a}{c}$ (B) $\dfrac{a}{b}$ (C) $\dfrac{b}{c}$ (D) $\dfrac{d}{a}$ (E) $\dfrac{-b}{d}$

35. Simplify $(\sqrt{2} - 3)(\sqrt{2} + 3)$.

(A) 3   (B) –7   (C) –3   (D) 5   (E) –5

36. Ten students are running in an event. In how many ways can three winners come in first, second and third places? Use the permutation formula $_nP_r = \dfrac{n!}{(n-r)!}$, where *n* is the number of items and *r* is the number of positions.

(A) 480   (B) 360   (C) 720   (D) 240   (E) 810

37. José has gotten 3 hits out of 5 times at bat. How many more hits does he need to raise his batting average to .800 (80%)?

(A) 5   (B) 6   (C) 3   (D) 2   (E) 4

38. At which point does the graph of the inverse of the function $y = 4^x$ cross the *x*-axis?

(A) (3, 0)   (B) (5, 1)   (C) (3, 1)   (D) (2, 0)   (E) (1, 0)

39. The sum of four consecutive odd integers is 32. What is their average?

(A) 4.9   (B) 6   (C) 8   (D) 5   (E) 7

40. Find the area of a triangle whose vertices on the co-ordinate plane axis are A(–3, 6), B(6, 2) and C(–3, –4).

(A) 45   (B) 50   (C) 35   (D) 50   (E) 38

41. Simplify

$$\sum_{i=2}^{5}(i+2)^2$$

(A) 154   (B) 126   (C) 76   (D) 88   (E) 132

42. If $\dfrac{f(x)}{x+3} = x - 2 +$ (remainder of 6), what is $f(x)$?

(A) $x^2 - 6x$ (B) $x^2 + 6x + 1$ (C) $x^2 - 2x + 6$ (D) $x^2 + 2x + 6$ (E) $x^2 + x$

Test 4

43. CD ⊥ FH, $m\angle CGI = 2x + 6$, $m\angle FGB = 3x - 16$. Find the measure of $\angle FGB$.

(A) 54°   (B) 44°   (C) 38°   (D) 52°   (E) 62°

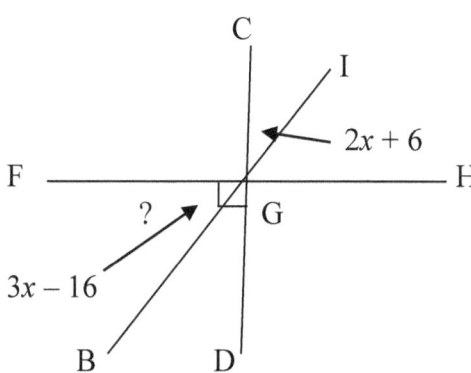

44. There are 2,400 students in a college. Twenty percent are seniors and forty percent of seniors are 5'9" or above. How many seniors are below 5'9"?

(A) 360   (B) 224   (C) 288   (D) 480   (E) 336

45. A factory produces widgets. The number of defective widgets is represented by $y\%$ of the total, t. Which fraction represents the number of good widgets compared to the total?

(A) $\dfrac{(y\%)t}{t}$ (B) $\dfrac{y\% - (y\%)t}{t}$ (C) $\dfrac{100\% - y\%}{t}$ (D) $\dfrac{(y\%)t}{(100\% - y\%)t}$

(E) $\dfrac{(100\% - y\%)t}{t}$

46. Two members of the debating team are to be chosen to represent the school in an area debate. There are 9 members on the team. What is the probability that either Quan or Latisha, two members of the team, will be chosen but not both?

(A) 17/81   (B) 2/9   (C) 15/27   (D) 18/27   (E) 15/81

47. If y is directly proportional to $\sqrt{x}$ and inversely proportional to z^2, find the positive value of x when $z = 5$, $y = 2$ and the constant of proportionality is 2.

(A) 430   (B) 225   (C) 250   (D) 62   (E) 475

48. The chart below indicates the number of miles driven by Max and by Jenny in 7 hours. What was Max's average speed for the first 5 hours?

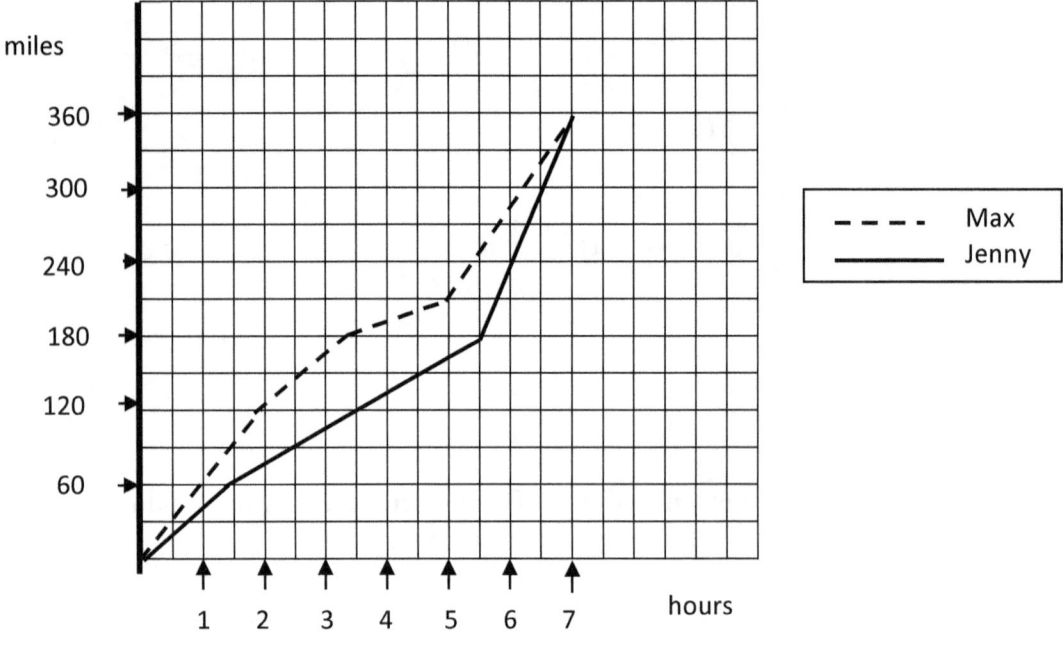

(A) 54 mph (B) 63 mph (C) 42 mph (D) 38 mph (E) 58 mph

49. Which laws hold for complex numbers?

(1) Commutative Property of Addition
(2) Commutative Property of Division
(3) Associative Property of Addition
(4) Associative Property of Multiplication
(5) Distributive Property

(A) 1, 2, 3 and 4   (B) 2, 3, 4, 5   (C) 1, 3, 5   (D) 1, 3, 4, 5   (E) 1, 2, 3, 4, 5

50. What is the locus of points 4 units away from the graph of the function $(x-2)^2 + (y-3)^2 = 4$?

(A) $x^2 + y^2 = 4^2$
(B) $(x-2)^2 + (y+3)^2 = 36$
(C) $(x-2)^2 + (y-3)^2 = 16$
(D) $(x-2)^2 + (y-3)^2 = 4$
(E) $(x-2)^2 + (y-3)^2 = 36$

51. Find the secant of angle A in the right triangle formed by the points indicated in the graph shown.

(A) 5/3 (B) $\frac{2}{\sqrt{3}}$ (C) 1/4 (D) 3/5 (E) 2/5

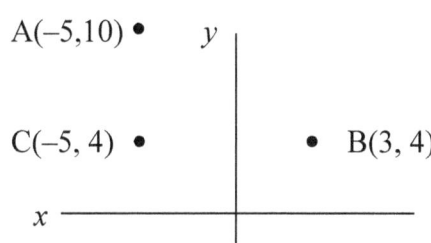

52. If tan A = 5/12, find cos 2A.

$$\cos(A + A) = \cos A \cdot \cos A - \sin A \cdot \sin A = \cos^2 A - \sin^2 A$$

(A) $\frac{119}{169}$ (B) $\frac{92}{118}$ (C) $\frac{57}{72}$ (D) $\frac{145}{189}$ (E) $\frac{35}{136}$

53. Find the value of *m* in the equation

$$\frac{m}{m-1} = \frac{2m-6}{2m}$$

(A) 1/2 (B) 5/8 (C) 7/8 (D) 3/4 (E) 2/3

Given *a* and *b* are integers, $-5 < a < -4$ and $0 < b < 3$, what is the largest approximate value for $a^2 + b^2$? Round off to the nearest integer.

(A) 36 (B) 34 (C) 37 (D) 32 (E) 33

55. Given that *x* and *y* are positive integers and $x > y$, which of the following choices is the best description of $\frac{|x-y|}{|x+y|}$.

(A) 1 (B) < 1 (C) > 1 (D) 2 (E) > 2

56. Water runs into a pool 9 meters long by 5 meters wide and 4 meters high at the rate of 120,000 cubic centimeters per minute. How long will it take to fill the pool?
l meter = 100 centimeters

(A) 25 hours (B) 46 hours (C) 30 hours (D) 50 hours (E) 38 hours

57. The gas tank of a car holds 25 gallons. The car uses one gallon per 22 miles driving on a horizontal road. It uses one gallon per 18 miles driving uphill. The driver has gone 462 miles on a horizontal road. For the rest of trip, it's uphill driving. How many more miles will the remainder of gas in the tank support driving on the uphill road?

(A) 84   (B) 72   (C) 104   (D) 98   (E) 85

58. Find the greatest prime number, x, that will satisfy the inequality.

$$2 \leq \sqrt{\frac{x+6}{3}} \leq \sqrt{47}$$

(A) 413   (B) 133   (C) 247   (D) 323   (E) 477

59. Select the set of equations to describe the area common to both in the figure shown.

(A) $y = 3 \cap y < x + 2$
(B) $y \geq 3 \cap y < x + 2$
(C) $y \leq 3 \cap y = x + 2$
(D) $y \leq x + 2 \cap y < 3$
(E) $y < 3 \cap y \leq x + 2$

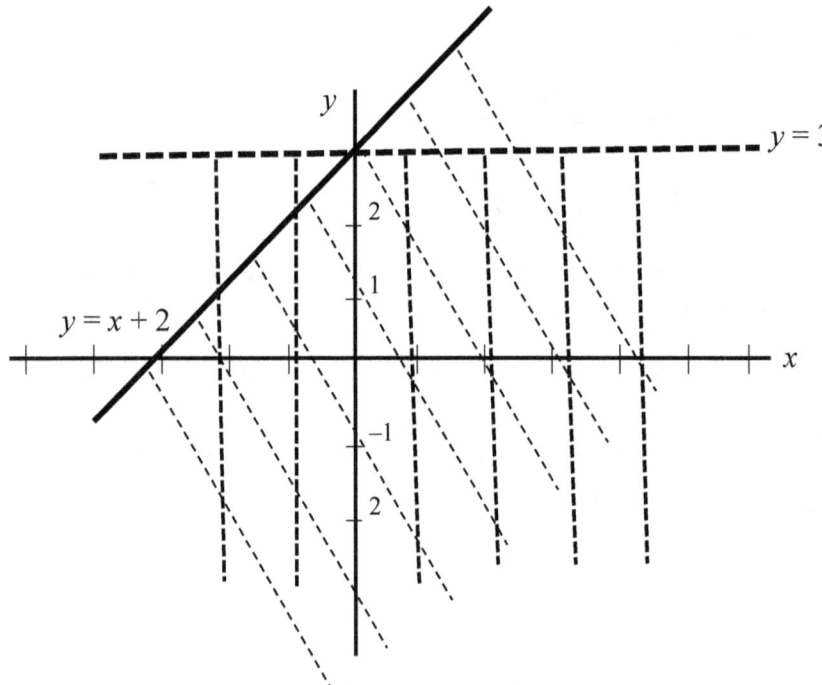

60. A metal pitcher filled with water weighs 21 ounces. If it's half full, the pitcher and the water weigh 12 ounces. How many ounces does the pitcher weigh alone?

(A) 5 ounces (B) 3 ounces (C) 6 ounces (D) 4 ounces (E) 2 ounces

ACT Math Personal Tutor

Answers to Test 4

1. Find the value of a in the equation $16^{a/2} = 8$.

(A) 2/3 (B) 1/3 (C) 3/4 (D) 3/2 (E) –2/3

$$16^{a/2} = 8$$

$$(2^4)^{a/2} = 2^3$$

$$2^{4a/2} = 2^3$$

$$2^{2a} = 2^3$$

Since the bases are now the same (2), equate the exponents:

$$2a = 3$$

$$a = 3/2$$

Answer: (D) 3/2

2. Let U = {a, b, c, d, e, f, g}, △ = {a, b, c, d}, ○ = {c, d, e, f}, and □ = {a}.
Find ~{□ ∩ ○} ∩ {○ ∩ ~△}.

(A) {b, c, d} (B) {d, e, f, g} (C) {e, f} (D) {a, b, c, d} (E) {d, e, f, g}

{□ ∩ ○} = [{a} ∩ {c, d, e, f}] = { }   That is, the empty set.

~{□ ∩ ○} = {a, b, c, d, e, f, g}   The opposite of the empty set is the universal set, U.

{○ ∩ ~△} = [{c, d, e, f} ∩ {e, f, g}] = {e, f}

~{□ ∩ ○} ∩ {○ ∩ ~△} = {a, b, c, d, e, f, g} ∩ {e, f} = {e, f}

Answer: (C) {e, f}

3. Subtract $5 - \sqrt{-32}$ from $3 + \sqrt{-18}$.

(A) $3 + 2i$ (B) $-2 + 7i\sqrt{2}$ (C) $5 - 4\sqrt{-2}$ (D) $3 - 4\sqrt{-3}$ (E) $3 + 5i\sqrt{3}$

Before performing any operations with complex numbers, first pull out the negative symbol from the radical and replace it with i^*.

$$\begin{array}{rcccl}
3 + \sqrt{-18} &=& 3 + \sqrt{-9 \cdot 2} &=& 3 + 3i\sqrt{2} \\
(-)\ 5 - \sqrt{-32} &=& 5 - \sqrt{-16 \cdot 2} &=& 5 - 4i\sqrt{2} \\
\hline
& & & & -2 + 7i\sqrt{2}
\end{array}$$

Answer: (B) $-2 + 7i\sqrt{2}$

4. Lucy purchases 100 shares of stock of ABC Corporation at $7 per share. The stock pays an annual dividend of 2.50% on the original purchase price of $7. The first year each share increased in value by 3%. The second year each share dropped 2% and the third year each share increased by 6%. Including dividends, how much is Lucy's investment worth at the end of 3 years? Select the closest answer.

(A) $909 (B) $854 (C) $746 (D) $917 (E) $802

Initial price	Price after 1 year	Price after 2 years	Price after 3 years	Three years of dividends	Total
$7	1.03 × $7 = $7.21	.98 × $7.21 = $7.07	1.06 × $7.07 = $7.49	.025 × 3 × $7 = $.53	$7.49 + $.53 = $8.02

100 shares × $8.02 = $802

Answer: (E) $802

5. C and D are different positive integers. Find the value of C in this addition problem.

```
   D C
 + C D
 -----
 1 C 2
```

(A) 2 (B) 1 (C) 7 (D) 3 (E) 8

Since C and D are *different* positive integers, they cannot both be 1 in order to add up to 2 in the right-hand column. Therefore, C + D = 12. Now, carry the 1 to the left-hand column, so C + D + 1 = 12 + 1 = 13. And C is in the third spot, so C = 3.

Answer: (D) 3

---

* Some properties of real numbers are not true for complex numbers. For example, $\sqrt{-2}\sqrt{-2} \neq \sqrt{(-2)(-2)} = \sqrt{4} = 2$. In fact, $\sqrt{-2}\sqrt{-2} = \sqrt{2}i \cdot \sqrt{2}i = 2i^2 = 2(-1) = -2$. Writing all complex numbers in terms i will avoid errors like this one.

6. Which one of the following points is on the graph of $f(x) = x^2 - x - 6 = 0$?

(A) (0, 1)   (B) (1, –6)   (C) (3, 0)   (D) (3, 4)   (E) (4, 5)

Try (C) (3, 0):
$$f(x) = x^2 - x - 6$$
$$0 = 3^2 - 3 - 6$$
$$0 = 0 \checkmark$$

Answer: (C) (3, 0)

7. Which equation is possibly represented by this graph?

(A) $y = x^2 + 2$ (B) $x - 2 = y$ (C) $y^2 = x + 1$ (D) $y = \pm\sqrt{x} + 2$ (E) $y = -x^2 + 2$

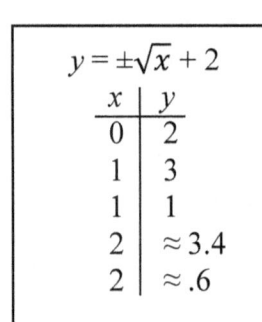
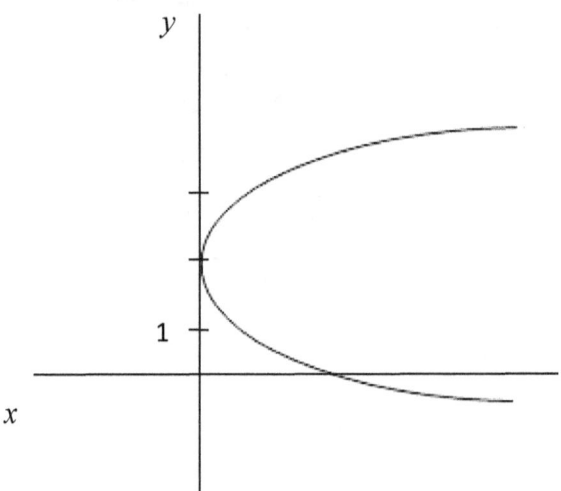

We can eliminate (A) and (E) because the graphs of these parabolas open facing up and down. We can also eliminate (B) because the graph is a straight line. You could use your graphing calculator to solve this problem.

(D) $y = \pm\sqrt{x} + 2$
$y - 2 = \pm\sqrt{x}$
$(y - 2)^2 = x$
$y^2 - 4y + 4 = x$
$x = y^2 - 4y + 4$

The equation simplifies to the form $x = ay^2 + by + c$, which indicates a sideways figure. The "2" in the original equation indicates that the vertex of the parabola is 2 units above the x-axis.

Answer: (D) $y = \pm\sqrt{x} + 2$

8. A harmonic sequence is a sequence of numbers whose reciprocals form an arithmetic sequence. If there are 5 terms in a harmonic sequence whose first term is 1/5 and whose last term is 1/21, what is the third term in the harmonic sequence?

(A) 1/13 (B) 3/11 (C) 1/7 (D) 2/7 (E) 2/11

Since 1/5 is the first term of the harmonic sequence, the first term of the corresponding arithmetic sequence is 5. Similarly, since 1/21 is the last term of the harmonic sequence, 21 is the last term of the arithmetic sequence.

Let's find the common difference in the arithmetic sequence. We'll use the formula $a_n = a_1 + (n-1)d$, where a_n = the last term, a_1 = the first term, n = the number of terms and d = the difference between successive terms:

$$a_n = a_1 + (n-1)d$$

$$21 = 5 + (5-1)d$$

$$21 = 5 + 4d$$

$$16 = 4d$$

$$d = 4$$

First term of the arithmetic sequence = 5
Second term of the arithmetic sequence = 5 + 4 = 9
Third term of the arithmetic sequence = 9 + 4 = 13
Fourth term of the arithmetic sequence = 13 + 4 = 17
Fifth term of the arithmetic sequence = 17 + 4 = 21

First term of the harmonic sequence = 1/5
Second term of the harmonic sequence = 1/9
Third term of the harmonic sequence = 1/13 ←
Fourth term of the harmonic seuqence = 1/17
Fifth term of the harmonic sequence = 1/21

Answer: (A) 1/13

9. Joel saved $20 for his Father's Day gift. His sister, Marilyn, saved 40% of the total amount of the gift. If they still needed $10, what was the total amount of the gift?

(A) $30   (B) $60   (C) $40   (D) $20   (E) $50

Joel's share = $20.
Let x = the total amount of the gift.
Marilyn's share = $.40x$

$$\text{Joel's share} + \text{Marilyn's share} + \$10 = \text{Total}$$
$$20 + .40x + 10 = x$$
$$30 + .40x = 1.00x$$
$$30 = .60x$$
$$x = \$50$$

Answer: (E) $50

10. Mr. Jenkins invests $9,000 at a 5% annual interest rate without compounding for 2 years. He then takes the total and invests it for 3 additional years. If he has $11,088 at the end of the 3 years and the interest rate was not compounded, what was the interest rate for the second investment?

(A) 3%   (B) 6%   (C) 4%   (D) 3.5%   (E) 4.5%

The first investment at the end of 2 years = initial investment + interest
$$= 9000 + .05(9000) \cdot 2 = 9900$$

Let x = the interest rate of the second investment.

The second investment at the end of 3 additional years = initial investment + interest
$$= 9900 + x(9900) \cdot 3$$

At the end of the second investment, Mr. Jenkins has $11,088, so

$$9900 + x(9900) \cdot 3 = 11{,}088$$
$$29{,}700x = 1{,}188$$
$$x = .04$$

Answer: (C) 4%

11. Multiply:

$$\frac{x^2 - x - 6}{2x - 6} \cdot \frac{4x - 12}{x^2 + 5x + 6}$$

(A) $\dfrac{2(x+2)}{x-3}$ (B) $\dfrac{3(x+3)}{x+2}$ (C) $\dfrac{x-3}{x+3}$ (D) $\dfrac{2(x-3)}{x+3}$ (E) $\dfrac{3(2x+1)}{x-2}$

$$\frac{x^2 - x - 6}{2x - 6} \cdot \frac{4x - 12}{x^2 + 5x + 6}$$

$$\frac{(x+2)(x-3)}{2(x-3)} \cdot \frac{4(x-3)}{(x+2)(x+3)}$$

$$\frac{2(x-3)}{x+3}$$

Answer: (D) $\dfrac{2(x-3)}{x+3}$

12. Below is a list of books borrowed in a month from the Gotham Library. The percents of the total number of books borrowed are listed, but only one category, biography, lists the actual number of books borrowed. How many history books were borrowed?

(A) 5,800 (B) 4,500 (C) 6,000 (D) 5,000 (E) 5,400

Books Borrowed from Gotham Library		
Category	**% of Total**	**Number**
mysteries	40	
poetry	5	
biography	20	4,000
history	25	
science	10	

Let x = the total number of books borrowed.

Biography lists 20%, so 20% of the total equals 4,000:

$$.20x = 4,000$$

Multiply by 100:
$$20x = 400,000$$
$$x = 20,000$$

Now, by the table, twenty-five percent were history books:

$$.25(20,000) = 5,000$$

Answer: (D) 5,000

13. Given the two equations $y = 2x - 5$ and $x = \dfrac{y}{3} + \dfrac{7}{3}$, where do their graphs intersect?

(A) (2, 1)   (B) (3, –1)   (C) (2, –1)   (D) (2, –3)   (E) (1, 4)

$$(1) \quad y = 2x - 5$$

Multiply the equation $x = \dfrac{y}{3} + \dfrac{7}{3}$ by 3:

$$(2) \quad 3x = y + 7$$

$$(2) \quad y = 3x - 7$$

Set the value of y equal in both equations (1) and (2):

$$2x - 5 = 3x - 7$$
$$x = 2$$

Substitute 2 for x in (1):
$$(1) \quad y = 2x - 5$$
$$(1) \quad y = 2(2) - 5$$
$$(1) \quad y = -1$$

Answer: (C) (2, –1)

14. If $a = 6r^2$ and the ratio of $a : b$ is $2 : 7$, what is the value of b in terms of r?

(A) $14r$ (B) $14r^2$ (C) $18r$ (D) $16r^2$ (E) $21r^2$

Writing the proportion $a : b$ as $2 : 7$ in terms of fractions gives

$$\frac{a}{b} = \frac{2}{7}$$

$a = 6r^2$:

$$\frac{6r^2}{b} = \frac{2}{7}$$

$$2b = 42r^2$$

$$b = 21r^2$$

Answer: (E) $21r^2$

15. There are a residents of Middle City above age 30 and b residents age 30 or under. If 600 children are born and 400 people above 30 leave the city, what fraction of the total new population represents the number of new residents above age 30?

(A) $\dfrac{a + 600}{a + b + 400}$ (B) $\dfrac{a - 400}{a - 600 + b}$ (C) $\dfrac{a + b - 600}{b + 400}$ (D) $\dfrac{a - 400}{a + b + 200}$

(E) $\dfrac{2a + b}{300 - a}$

Original population above age 30: a
Original population age 30 or under: b

New population above age 30 =

(old population above age 30) − (400 residents above age 30 who leave the city) =

$a - 400$

New population age 30 or below =

(old population 30 or below) + (newborn children) =

$b + 600$

Total new population: $(a - 400) + (b + 600) = a + b + 200$

$$\dfrac{\text{New population above age 30}}{\text{Total new population}} = \dfrac{a - 400}{a + b + 200}$$

Answer: (D) $\dfrac{a - 400}{a + b + 200}$

16. If a possible span of the area, *A*, of a semi-circle is $12.5\pi \leq A \leq 18\pi$, what is a possible span of values for the radius, *r*?

(A) $5 \leq r \leq 6$ (B) $3 < r < 4$ (C) $4 \; r < 5$ (D) $2 \leq r \leq 3$ (E) $5 < r < 7$

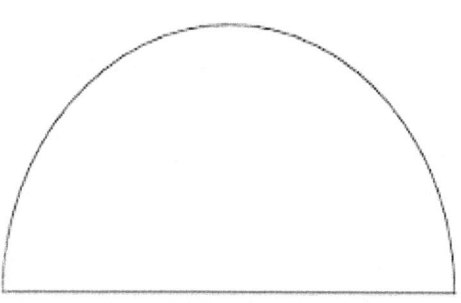

If $r = 2$, $A = \frac{1}{2}\pi(2)^2 = 2\pi$ – too small an area.

If $r = 3$, $A = \frac{1}{2}\pi(3)^2 = 4.5\pi$ – too small an area.

If $r = 4$, $A = \frac{1}{2}\pi(4)^2 = 8\pi$ – too small an area.

If $r = 5$, $A = \frac{1}{2}\pi(5)^2 = 12.5\pi$ – OK area.

If $r = 6$, $A = \frac{1}{2}\pi(6)^2 = 18\pi$ – OK area.

If $r = 7$, $A = \frac{1}{2}\pi(7)^2 = 24.5\pi$ – too large an area.

Area of a circle = πr^2
Area of a semicircle = $\frac{1}{2}\pi r^2$

Answer: (A) $5 \leq r \leq 6$

17. The units' digit of a three-digit number is 8. The hundreds' digit is 3. Find the tens' digit if a multiple of the number is 1014.

(A) 2 (B) 3 (C) 4 (D) 5 (E) 6

The number has the following structure:

$$3\_8$$

Here, the underscore indicates the missing tens' digit.

Let's just substitute the digits 0, 1, 2, 3, ..., 9 into the tens' digit position in 3 _ 8 until we get the number that is a factor of 1014:

1 · 308 = 308	1 · 318 = 318	1 · 328 = 328	1 · 338 = 338
2 · 308 = 612	2 · 318 = 636	2 · 328 = 656	2 · 338 = 676
3 · 308 = 924	3 · 318 = 954	3 · 328 = 984	**3 · 338 = 1014**

Answer: (B) 3

18. Two points, (4, 5) and (2, 7), lie on the circumference of a circle at the ends of a diameter. Find the center of the circle.

(A) (2, 4) (B) (3, 5) (C) (4, 6) (D) (3, 4) (E) (3, 6)

We can find the center of the circle by finding the midpoint of the line connecting (4, 5) and (2, 7).

The midpoint between (4, 5) and (2, 7) is

$$\left(\frac{4+2}{2}, \frac{5+7}{2}\right) = (3, 6)$$

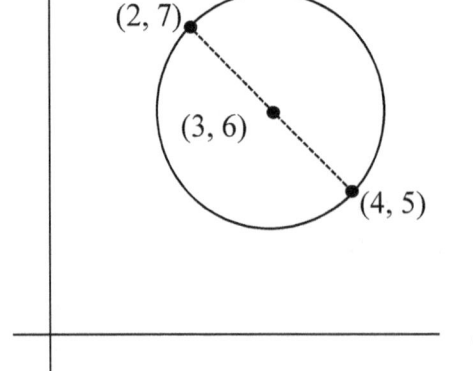

Answer: (E) (3, 6)

19. In the figure, triangle ABC is isosceles and AB ≅ BC. DE ∥ FG ∥ AC. $m\angle BHG = 115°$. Find the measure of ∠B.

(A) 50°   (B) 30°   (C) 40°   (D) 60°   (E) 70°

$m\angle GHC = 180° - 115° = 75°$
$m\angle C = m\angle GHC = 75°$ because alternate interior angles of parallel lines are congruent
$m\angle C = m\angle A = 75°$ because base angles of an isosceles triangle are congruent
$m\angle B = 180° - 75° - 75° = 30°$ because the measures of the angles of a triangle sum to 180°

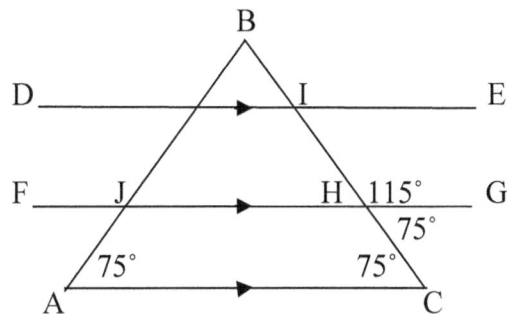

Answer: (B) 30°

20. Given horizontal line l, how many points are located 4 inches away from that line and 5 inches away from point a on that line?

(A) 1   (B) 3   (C) 4   (D) 2   (E) none

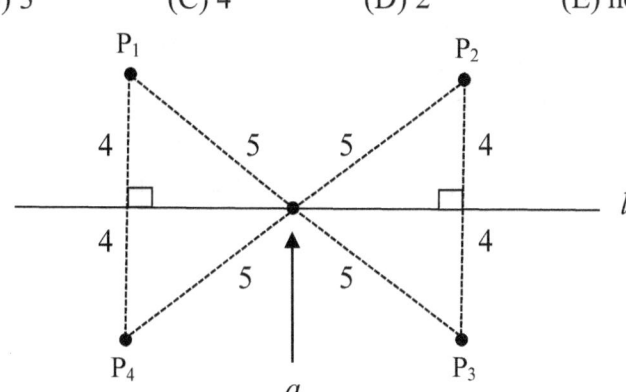

Answer: (C) 4

21. Diamond Airlines flies twice as many passengers per day as Euclid Air. Euclid flies 1/3 as many passengers per day as Maloney Air. If the three airlines together fly 30,000 passengers per day, how many passengers does Maloney Air fly per day?

(A) 20,000   (B) 15,000   (C) 10,000   (D) 12,000   (E) 18,000

If Euclid flies 1/3 as many passengers as Maloney and Euclid flies x passengers, Maloney flies $3x$ passengers.

Diamond = $2x$
Euclid = x
Maloney = $3x$

$$2x + x + 3x = 30,000$$
$$6x = 30,000$$
$$x = 5,000$$
$$2x = 10,000$$
$$3x = 15,000$$

Answer: (B) 15,000

22. If $90° < \angle B < 180°$ and $\sin B = 1/2$, find $\sec B$.

(A) $\dfrac{\sqrt{3}}{2}$ (B) $\dfrac{1}{\sqrt{3}}$ (C) 2   (D) $\dfrac{-2}{\sqrt{3}}$ (E) $\dfrac{-\sqrt{3}}{2}$

Applying the Pythagorean Theorem to the right triangle in the figure gives

$a^2 + b^2 = c^2$
$1^2 + x^2 = 2^2$
$x^2 = 3$
$x = \sqrt{3}$

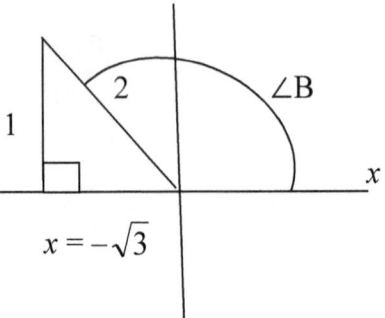

$x = -\sqrt{3}$

$$\sec B = \dfrac{1}{\cos B} = \dfrac{1}{-\sqrt{3}/2} = \dfrac{-2}{\sqrt{3}}$$

Answer: (D) $\dfrac{-2}{\sqrt{3}}$

23. Find the area of the triangle in the diagram shown.

(A) $de + ef$ (B) $de + e$ (C) $fe + \frac{1}{2}d$ (D) $df + e$ (E) $de + f$

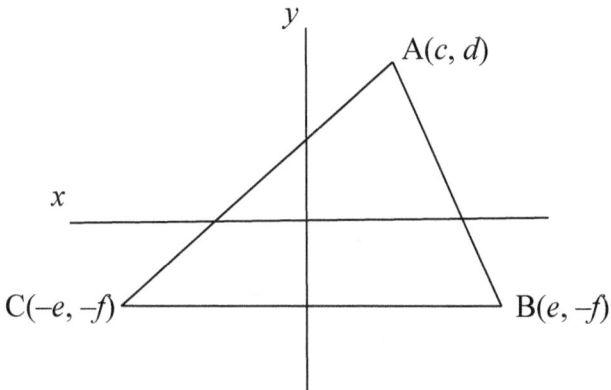

$A = \frac{1}{2}bh$, where A = area, b = base, h = height

$b = e - (-e) = 2e$
$h = d - (-f) = d + f$

$$A = \frac{1}{2}(2e)(d+f) =$$
$$e(d+f) =$$
$$de + ef$$

Answer: (A) $de + ef$

24. If x = arctangent (-2.5), where is x located?

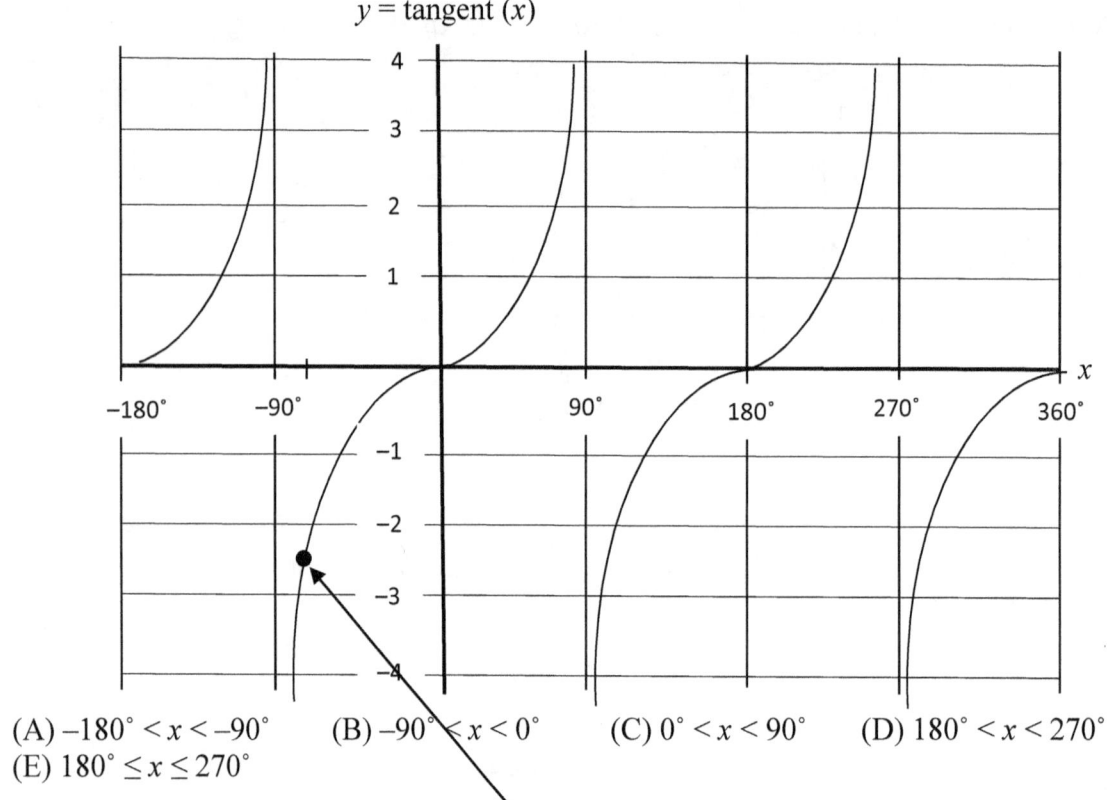

(A) $-180° < x < -90°$ (B) $-90° < x < 0°$ (C) $0° < x < 90°$ (D) $180° < x < 270°$
(E) $180° \leq x \leq 270°$

The arctangent (-2.5) is the angle for which the tangent function has a value of -2.5. That is,

$$\tan x = -2.5$$

Because the tangent function is periodic, it takes on the value -2.5 repeatedly. But only one interval $(-90° < x < 0°)$ is offered where $\tan x = -2.5$.

Answer: (B) $-90° < x < 0°$

Test 4

25. The XYZ Corporation manufactures auto tires. The manufacturing cost of each tire, C, varies jointly as the cost of rubber per tire, R, and the cost of labor per tire, L. The cost of manufacturing a tire is $18 when the labor is $4 and rubber $3. Find the cost of manufacturing one tire if labor goes up to $5 and rubber drops to $2.

(A) $15   (B) $16   (C) $17   (D) $18   (E) $19

The definition of *joint variation* is if y varies jointly with x and z, then $y = kxz$, where k is a constant.

Step 1 (find the constant k): $C = kRL$, where k is a constant.
$C = 18, R = 3, L = 4$:

$$18 = k(3)(4)$$
$$18 = 12k$$
$$k = 1.5$$

Step 2 (evaluate the function C): $C = 1.5RL$.
$R = 2, L = 5$:

$$C = 1.5(2)(5)$$
$$C = 15$$

Answer: (A) $15

26. Jenny has to register for 3 college courses. She can take 1 out of 6 math classes, 1 out of 4 history classes, and 2 out of 5 science classes. How many different combinations of 3 courses can she choose?

(A) 480   (B) 120   (C) 240   (D) 180   (E) 600

math   history   science
any of 6   any of 4   2 out of 5

6   ×   4   ×   5 · 2   = 240

Answer: (A) 480

27. What is the maximum point of the graph of the function $f(x) = -x^2 - 6x + 5$?

(A) (3, 10)   (B) (–3, 14)   (C) (6, 10)   (D) (7, 12)   (E) (2, 12)

In the general form of the quadratic equation $y = ax^2 + bx + c$, if $a < 0$, the graph has a maximum point at $x = \dfrac{-b}{2a}$.

$y = -x^2 - 6x + 5$

$x = \dfrac{-b}{2a} = \dfrac{-(-6)}{2(-1)} = \dfrac{6}{-2} = -3$

To get the y-value, substitute $x = -3$ into $f(x)$.

$f(-3) = -(-3)^2 - 6(-3) + 5 = 14$

The vertex is (3, 14).

Answer: (B) (–3, 14)

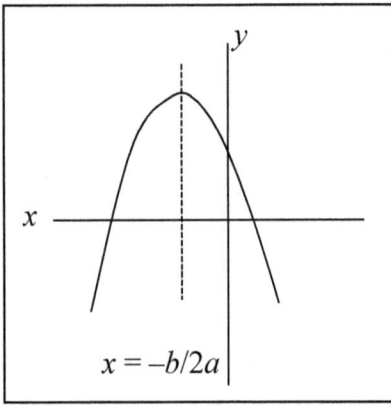

28. The letters *a*, *b* and *c* represent positive integers such that $2c = 3a$ and $-4b = c$. Find the value of *a* when $a + b + c = 3$.

(A) $2\frac{3}{11}$ (B) $3\frac{4}{9}$ (C) $3\frac{8}{11}$ (D) $3\frac{5}{7}$ (E) $1\frac{7}{17}$

Solving the equation $2c = 3a$ for *c* yields

$$(1) \quad c = \frac{3a}{2}$$

Solving the equation $-4b = c$ for *b* yields

$$(2) \quad b = \frac{-c}{4}$$

Substituting equation (1) into equation (2) yields

$$(3) \quad b = \frac{-3a/2}{4} = \frac{-3a}{2} \cdot \frac{1}{4} = \frac{-3a}{8}$$

Now, plug equations (1) and (3) into the equation $a + b + c = 3$ to express it in terms of only the variable *a*:

$$a + \frac{-3a}{8} + \frac{3a}{2} = 3$$

Multiply by 8 to clear the fractions:

$$8a - 3a + 12a = 24$$

$$17a = 24$$

$$a = \frac{24}{17} = 1\frac{7}{17}$$

Answer: (E) $1\frac{7}{17}$

29. In an analog clock, which is functioning exactly, how many degrees are there between the hour hand and the minute hand at 9:26?

(A) 116° (B) 138° (C) 127° (D) 135° (E) 142°

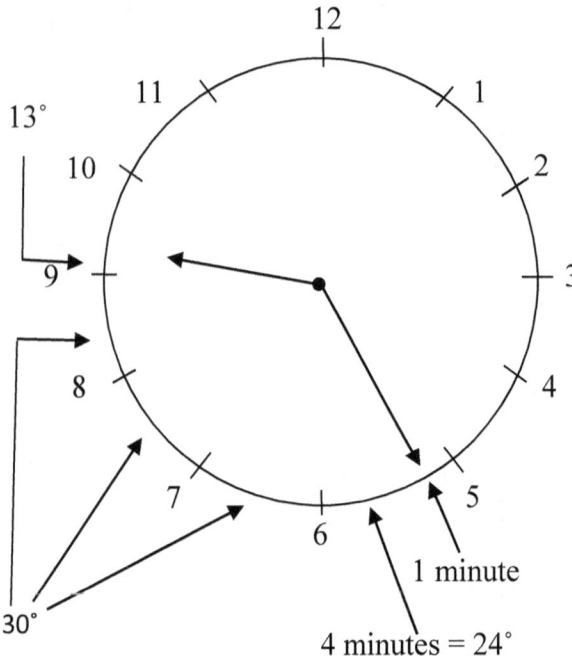

There are 360° in a circle and there are 60 minutes in an hour, so 360°/60 = 6° per minute.

There are 5 minutes × 6° = 30° between each number (or hour).

Degrees between 6 and 7 = 30°
Degrees between 7 and 8 = 30°
Degrees between 8 and 9 = 30°

The hour hand is between 9 and 10. 26 minutes is $\frac{26}{60}$ = 13/30 part of an hour. So, $\frac{13}{30}$ × 30° = 13°

The minute hand is between 5 and 6. At 26 minutes, the minute hand is 1 minute beyond the 5 but 4 minutes between the minute hand and 6. 4 minutes = 4/5 of the degrees between 5 and 6. So, $\frac{4}{5}$ × 30° = 24°.

Let's total all the degrees: 30°
                              30°
                              30°
                              13°
                           + 24°
                             127°

Answer: (C) 127°

30. Select the graph which represents the answer for the inequality $x^2 - 2x - 8 \geq 0$.

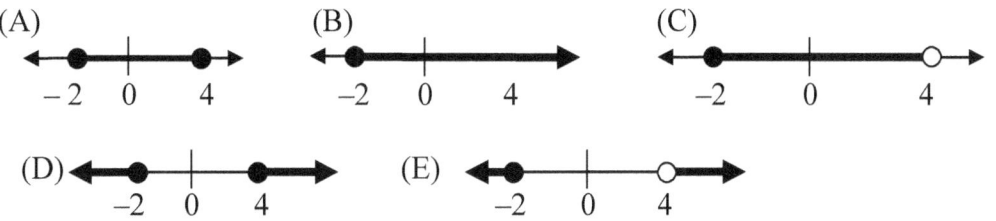

Factor to find the roots.

$$x^2 - 2x - 8 \geq 0$$

$$(x + 2)(x - 4) \geq 0$$

$$x = -2 \text{ and } x = 4$$

Graph it:

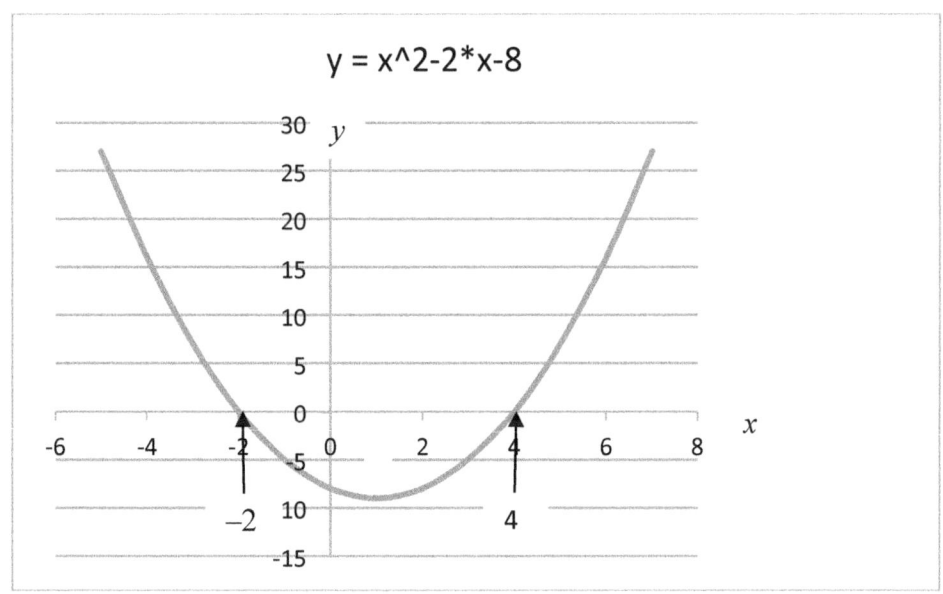

According to the graph, any y value greater than or equal to 4 or any y value less than or equal to –2 will satisfy the inequality $x^2 - 2x - 8 \geq 0$.

Answer: (D)

ACT Math Personal Tutor

31. Jocelyn makes a deposit of $1,000 in a bank account. If the bank pays a rate of 2% a year and the deposit is compounded four times a year, how much will she have in her account at the end of six years? Use the formula

$$A = P\left(1 + \frac{r}{n}\right)^{nt}$$

where A = the final amount in the account, P = the initial principle, r = the annual rate, t = the number of years involved and n = the number of times a year compounding takes place.

(A) $1,000(1.05)^4$ (B) $1,000(1.05)^{24}$ (C) $1,000(1.005)^{10}$
(D) $1,000(1.005)^{24}$ (E) $1,000(24)^{.05}$

$$A = P\left(1 + \frac{r}{n}\right)^{nt}$$

$$= 1000\left(1 + \frac{.02}{4}\right)^{4 \cdot 6}$$

$$= 1000(1 + .005)^{24}$$

$$= 1000(1.005)^{24}$$

Answer: (D) $1,000(1.005)^{24}$

32. In the function shown, if $f(k) = 2$, then which one of the following could be a value of k?

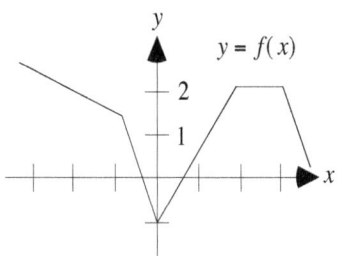

(A) –1   (B) 0   (C) 0.5   (D) 2.5   (E) 4

The graph has a height of 2 for every value of x between 2 and 3; it also has a height of 2 at about $x = -2$. The only number offered in this interval is 2.5. This is illustrated by the dot and the thick line in the following graph:

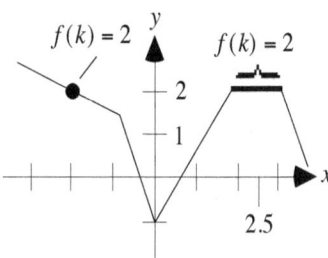

Answer: (D) 2.5

33. Simplify

$$\frac{\csc \theta}{\sin \theta} \cdot (1 - \cos^2 \theta)$$

(A) 1 (B) 2 (C) $\sin \theta$ (D) 3 (E) $\csc^2 \theta$

Some basic trigonometric identities are:

$$\csc \theta = \frac{1}{\sin \theta} \qquad \sec \theta = \frac{1}{\cos \theta} \qquad \cot \theta = \frac{1}{\tan \theta} \qquad \sin^2 \theta + \cos^2 \theta = 1$$

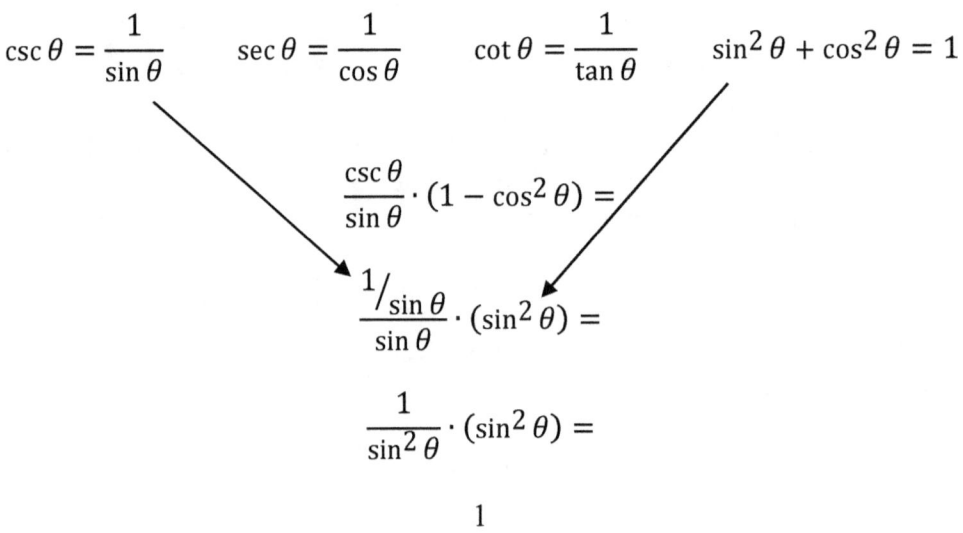

$$\frac{\csc \theta}{\sin \theta} \cdot (1 - \cos^2 \theta) =$$

$$\frac{1/\sin \theta}{\sin \theta} \cdot (\sin^2 \theta) =$$

$$\frac{1}{\sin^2 \theta} \cdot (\sin^2 \theta) =$$

$$1$$

Answer: (A) 1

34. The letters *a*, *b*, *c* and *d* represent integers in ascending order of size. The letters *a* and *b* represent negative integers, while *c* and *d* represent positive integers. Which choice is the smallest?

(A) $\dfrac{a}{c}$ (B) $\dfrac{a}{b}$ (C) $\dfrac{b}{c}$ (D) $\dfrac{d}{a}$ (E) $\dfrac{-b}{d}$

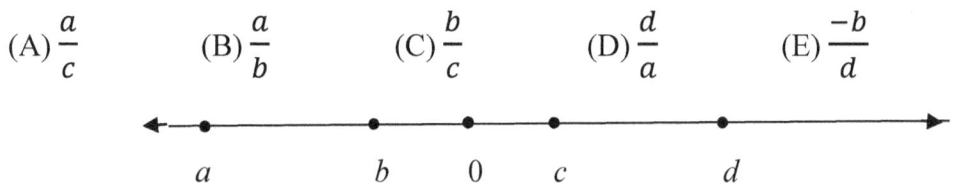

The easiest way to solve this problem is to arbitrarily assign integer values to the letters.

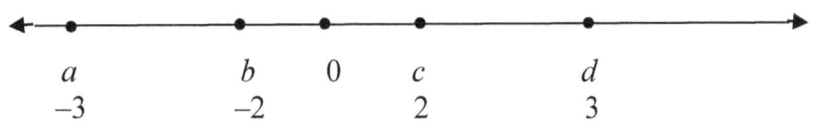

(A) $\dfrac{a}{c} = \dfrac{-3}{2} = -1.5$

(B) $\dfrac{a}{b} = \dfrac{-3}{-2} = +1.5$

(C) $\dfrac{b}{c} = \dfrac{-2}{2} = -1$

(D) $\dfrac{d}{a} = \dfrac{3}{-3} = -1$

(E) $\dfrac{-b}{d} = \dfrac{2}{3}$

Answer: (A) $\dfrac{a}{c}$

35. Simplify $(\sqrt{2} - 3)(\sqrt{2} + 3)$.

(A) 3 (B) –7 (C) –3 (D) 5 (E) –5

Use FOIL multiplication:

$$(\sqrt{2} - 3)(\sqrt{2} + 3) = 2 + 3\sqrt{2} - 3\sqrt{2} - 9 = -7$$

Answer: (B) –7

Method II (Difference of Squares Formula):

$$(\sqrt{2} - 3)(\sqrt{2} + 3) = (\sqrt{2})^2 - 3^2 = 2 - 9 = -7$$

36. Ten students are running in an event. In how many ways can three winners come in first, second and third places? Use the permutation formula $_nP_r = \dfrac{n!}{(n-r)!}$, where n is the number of items and r is the number of positions.

(A) 480 (B) 360 (C) 720 (D) 240 (E) 810

This is a permutation. Order does count.

So, $_{10}P_3 = \dfrac{10!}{(10-7)!} = \dfrac{10!}{7!} = \dfrac{10 \cdot 9 \cdot 8 \cdot \cancel{7 \cdot 6 \cdot 5 \cdot 4 \cdot 3 \cdot 2 \cdot 1}}{\cancel{7 \cdot 6 \cdot 5 \cdot 4 \cdot 3 \cdot 2 \cdot 1}} = 720$

Answer: (C) 720

37. José has gotten 3 hits out of 5 times at bat. How many more hits does he need to raise his batting average to .800 (80%)?

(A) 5 (B) 6 (C) 3 (D) 2 (E) 4

Let x = number of new hits.

$$\dfrac{\text{Number of old hits} + x}{\text{Total number of times at bat}} =$$

$$\dfrac{3 + x}{5 + x} = .80$$

Multiply by $5 + x$:

$$3 + x = .80(5 + x)$$
$$3 + x = 4 + .8x$$
$$.2x = 1$$
$$x = 5$$

Answer: (A) 5

38. At which point does the graph of the inverse of the function $y = 4^x$ cross the x-axis?

(A) (3, 0)   (B) (5, 1)   (C) (3, 1)   (D) (2, 0)   (E) (1, 0)

$$y = 4^x$$

Forming the inverse function by interchanging the x and y variables yields

$$x = 4^y$$

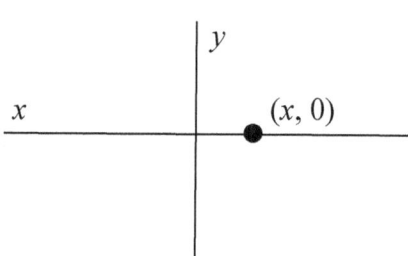

The function $x = 4^y$ crosses the x-axis when $y = 0$:

$$x = 4^0$$
$$x = 1$$

Answer: (E) (1, 0)

39. The sum of four consecutive odd integers is 32. What is their average?

(A) 4.9   (B) 6   (C) 8   (D) 5   (E) 7

Let x = the first odd integer.

Then $x + 2$ = the second odd integer, $x + 4$ = the third odd integer and $x + 6$ = the fourth odd integer.

$$x + (x + 2) + (x + 4) + (x + 6) = 32$$
$$4x + 12 = 32$$
$$4x = 20$$

$$\begin{aligned} x &= 5 \\ x + 2 &= 7 \\ x + 4 &= 9 \\ (+) \quad x + 6 &= 11 \\ \hline & 32 \end{aligned}$$

$$Average = \frac{Sum\ of\ the\ numbers}{Number\ of\ numbers} = \frac{32}{4} = 8$$

Answer: (C) 8

Method II (Shortcut):

We are given four numbers and their sum, 32. Hence, the average is 32/4 = 8.

40. Find the area of a triangle whose vertices on the co-ordinate plane axis are A(–3, 6), B(6, 2) and C(–3, –4).

(A) 45 (B) 50 (C) 35 (D) 50 (E) 38

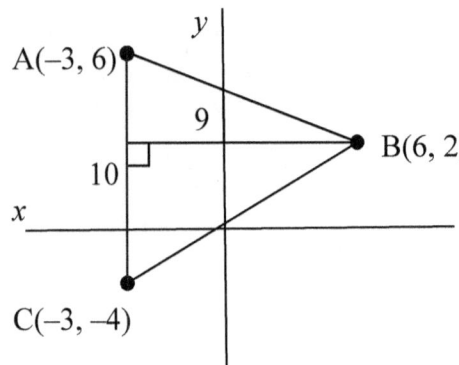

base (b) = 6 – (–4) = 10, height (h) = 6 – (–3) = 9

$$\text{Area} = \frac{1}{2}bh$$
$$= \frac{1}{2}(10)(9)$$
$$= 45$$

Answer: (A) 45

41. Simplify

$$\sum_{i=2}^{5}(i+2)^2$$

(A) 154 (B) 126 (C) 76 (D) 88 (E) 132

$$\sum_{i=2}^{5}(i+2)^2 = (2+2)^2 + (3+2)^2 + (4+2)^2 + (5+2)^2$$
$$\phantom{\sum_{i=2}^{5}(i+2)^2 =\;} i=2 \quad\quad i=3 \quad\quad i=4 \quad\quad i=5$$

$$= 4^2 + 5^2 + 6^2 + 7^2$$

$$= 16 + 25 + 36 + 49$$

$$= 126$$

Answer: (B) 126

42. If $\dfrac{f(x)}{x+3} = x - 2 +$ (remainder of 6), what is $f(x)$?

(A) $x^2 - 6x$ (B) $x^2 + 6x + 1$ (C) $x^2 - 2x + 6$ (D) $x^2 + 2x + 6$ (E) $x^2 + x$

Let's review some of the properties of division. Saying 7 divided by 3 equals 2 with a remainder of 1 means that 7 can be written as a product of 3 and 2 plus the remainder 1:

$$7 = 3 \cdot 2 + 1$$

Now, we are given that $f(x)$ divided by $x + 3$ equals $x - 2$ with a remainder of 6. This means that $f(x)$ can be written as a product of $x + 3$ and $x - 2$ plus the remainder 6:

$$f(x) = (x + 3)(x - 2) + 6 =$$

$$(x^2 + x - 6) + 6 =$$

$$x^2 + x$$

Answer: (E) $x^2 + x$

43. CD ⊥ FH, $m\angle CGI = 2x + 6$, $m\angle FGB = 3x - 16$. Find the measure of $\angle FGB$.

(A) 54° (B) 44° (C) 38° (D) 52° (E) 62°

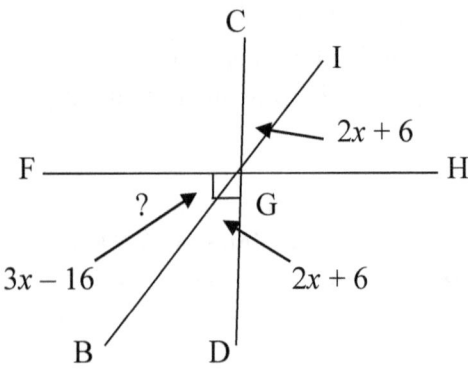

$m\angle BGD = m\angle CGI = 2x + 6$ because vertical angles are equal.

$m\angle FGD = 90°$ because CD ⊥ FH

$$(3x - 16) + (2x + 6) = 90°$$
$$5x = 100°$$
$$x = 20°$$

$$m\angle FGB = 3x - 16$$
$$= 3(20) - 16$$
$$= 44°$$

Answer: (B) 44°

44. There are 2,400 students in a college. Twenty percent are seniors and forty percent of seniors are 5'9" or above. How many seniors are below 5'9"?

(A) 360 (B) 224 (C) 288 (D) 480 (E) 336

Seniors = .20 × 2,400 = 480

Since 40% of seniors are 5 ft 9 in or above, 60% are below that height.

Seniors below 5 ft 9 in = .60 × 480 = 288

Answer: (C) 288

45. A factory produces widgets. The number of defective widgets is represented by $y\%$ of the total, t. Which fraction represents the number of good widgets compared to the total?

(A) $\dfrac{(y\%)t}{t}$ (B) $\dfrac{y\% - (y\%)t}{t}$ (C) $\dfrac{100\% - y\%}{t}$ (D) $\dfrac{(y\%)t}{(100\% - y\%)t}$

(E) $\dfrac{(100\% - y\%)t}{t}$

t = total number of widgets produced
$y\%$ are defective
$(100\% - y\%)t$ = percentage of good widgets

$$\frac{Good\ widgets}{Total\ widgets} = \frac{(100\% - y\%)t}{t}$$

Answer: (E) $\dfrac{(100\% - y\%)t}{t}$

46. Two members of the debating team are to be chosen to represent the school in an area debate. There are 9 members on the team. What is the probability that either Quan or Latisha, two members of the team, will be chosen but not both?

(A) 17/81 (B) 2/9 (C) 15/27 (D) 18/27 (E) 15/81

Probability of Quan: $\dfrac{1}{9}$
Probability of Latisha: $\dfrac{1}{9}$
Probability of both: $\dfrac{1}{9} \cdot \dfrac{1}{9} = \dfrac{1}{81}$

Probability of either one being chosen

$$\overbrace{\left(\frac{1}{9} + \frac{1}{9}\right) - \underbrace{\left(\frac{1}{9} \cdot \frac{1}{9}\right)}_{\text{Probability of both being chosen}}} = \frac{2}{9} - \frac{1}{81} = \frac{17}{81}$$

Answer: (A) 17/81

47. If y is directly proportional to $\sqrt{x}$ and inversely proportional to z^2, find the positive value of x when $z = 5$, $y = 2$ and the constant of proportionality is 2.

(A) 430 (B) 225 (C) 250 (D) 62 (E) 475

Translating "y is directly proportional to $\sqrt{x}$ and inversely proportional to z^2" into an equation gives

$$y = \frac{k\sqrt{x}}{z^2}$$

Plugging in the given information ($y = 2$, $k = 2$, $z = 5$) gives

$$2 = \frac{2\sqrt{x}}{5^2}$$

$$2 = \frac{2\sqrt{x}}{25}$$

$$50 = 2\sqrt{x}$$

$$25 = \sqrt{x}$$

$$625 = x$$

Answer: (D) 625

48. The chart below indicates the number of miles driven by Max and by Jenny in 7 hours. What was Max's average speed for the first 5 hours?

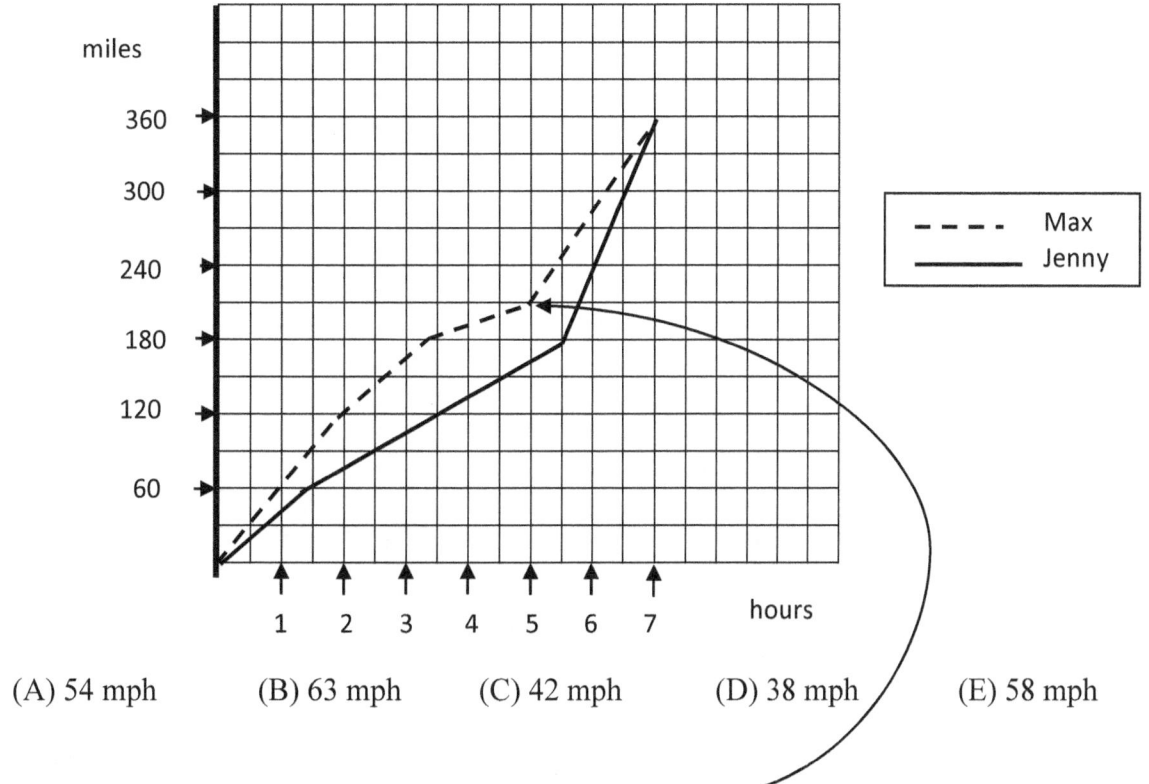

(A) 54 mph (B) 63 mph (C) 42 mph (D) 38 mph (E) 58 mph

From the chart, after 5 hours, Max has driven 210 miles.

$$Average\ Speed = \frac{Total\ Distance}{Total\ Time} = \frac{210}{5} = 42\ mph$$

Answer: (C) 42 mph

49. Which laws hold for complex numbers?

(1) Commutative Property of Addition
(2) Commutative Property of Division
(3) Associative Property of Addition
(4) Associative Property of Multiplication
(5) Distributive Property

(A) 1, 2, 3 and 4   (B) 2, 3, 4, 5   (C) 1, 3, 5   (D) 1, 3, 4, 5   (E) 1, 2, 3, 4, 5

Our strategy here is to map the problem back into the real number system, then use the properties of real numbers, and then map the problem back into the complex number system.

Let's use $a + bi$, $c + di$, and $e + fi$ to represent the three complex numbers that we will use to verify or disprove the 5 given properties.

For property (1), let's add the two complex numbers $a + bi$ and $c + di$:

$$(a + bi) + (c + di)$$

By definition, to add two complex numbers, add the real parts and the imangiary parts:

$$(a + c) + (b + d)i$$

Notice that $a + c$ and $b + d$ have been separated from the complex part i, so they are real numbers. Hence, we can apply the commutative property of *real* numbers to them:

$$a + c = c + a$$
$$b + d = d + b$$

Plugging these results back into $(a + c) + (b + d)i$ yields

$$(c + a) + (d + b)i$$
$$c + a + di + bi \qquad \text{by distributing the } i$$
$$c + di + a + bi$$
$$(c + di) + (a + bi)$$

Hence, Property (1) is true. Using this method, we can prove and disprove the remaining properties:

(2) $(a + bi) / (c + di) \neq (c + di) / (a + bi)$ ✗
(3) $\{(a + bi) + (c + di)\} + \{e + fi\} = \{a + bi\} + \{(c + di) + (e + fi)\}$ ✓
(4) $\{(a + bi) \cdot (c + di)\} \cdot \{e + fi\} = \{a + bi\} \cdot \{(c + di) \cdot (e + fi)\}$ ✓
(5) $\{a + bi\} \cdot \{(c + di) + (e + fi)\} = \{a + bi\} \cdot \{c + di\} + \{a + bi\} \cdot \{e + fi\}$ ✓

Answer: (D) 1, 3, 4, 5

50. What is the locus of points 4 units away from the graph of the function $(x-2)^2 + (y-3)^2 = 4$?

(A) $x^2 + y^2 = 4^2$
(B) $(x-2)^2 + (y+3)^2 = 36$
(C) $(x-2)^2 + (y-3)^2 = 16$
(D) $(x-2)^2 + (y-3)^2 = 4$
(E) $(x-2)^2 + (y-3)^2 = 36$

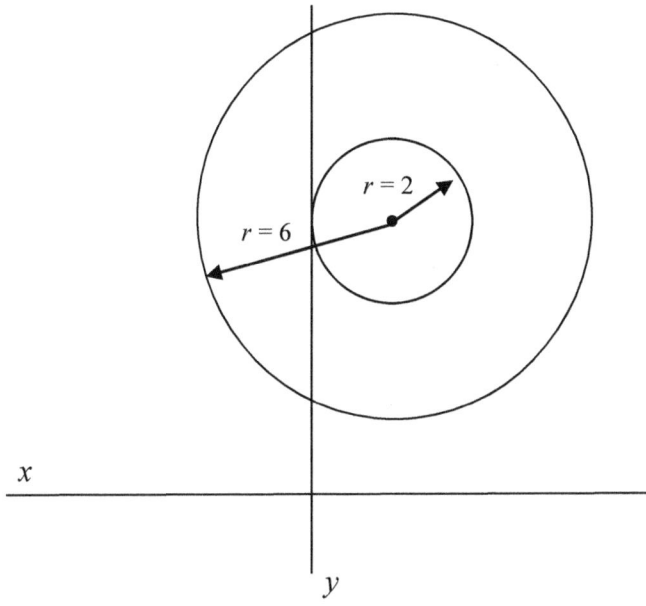

The smaller circle is the graph of the equation $(x-2)^2 + (y-3)^2 = 4$, where the center of the circle is at (2, 3).

The locus of points 4 units away from the smaller circle is the graph of the equation $(x-2)^2 + (y-3)^2 = 36$, where the radius is 6.

Answer: (E) $(x-2)^2 + (y-3)^2 = 36$

51. Find the secant of angle A in the right triangle formed by the points indicated in the graph shown.

(A) 5/3   (B) $\dfrac{2}{\sqrt{3}}$ (C) 1/4   (D) 3/5   (E) 2/5

From the figure, the length of the line AC is 6 (= 10 − 4). Let's use the distance formula to calculate the length of the line AB:

$$AB = \sqrt{(x_2 - x_1)^2 + (y_2 - y_1)^2} = \sqrt{(-5 - 3)^2 + (10 - 4)^2}$$
$$= \sqrt{(-8)^2 + (6)^2} = \sqrt{64 + 36} = \sqrt{100} = 10$$

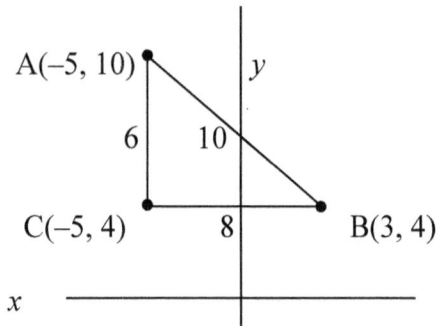

$$\sec A = \dfrac{1}{\cos A}$$
$$\cos A = \dfrac{6}{10} = \dfrac{3}{5}$$
$$\sec A = \dfrac{5}{3}$$

Answer: (A) 5/3

52. If tan A = 5/12, find cos 2A.

$$\cos(A + A) = \cos A \cdot \cos A - \sin A \cdot \sin A = \cos^2 A - \sin^2 A$$

(A) $\dfrac{119}{169}$ (B) $\dfrac{92}{118}$ (C) $\dfrac{57}{72}$ (D) $\dfrac{145}{189}$ (E) $\dfrac{35}{136}$

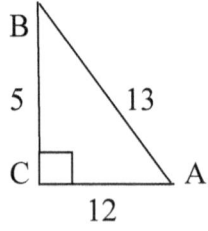

$$a^2 + b^2 = c^2$$
$$5^2 + 12^2 = c^2$$
$$c^2 = 169$$
$$c = 13$$

$$\cos 2A = \cos(A + A) = \cos A \cdot \cos A - \sin A \cdot \sin A$$
$$= \dfrac{12}{13} \cdot \dfrac{12}{13} - \dfrac{5}{13} \cdot \dfrac{5}{13}$$
$$= \dfrac{144}{169} - \dfrac{25}{169}$$
$$= \dfrac{119}{169}$$

Answer: (A) $\dfrac{119}{169}$

53. Find the value of m in the equation

$$\frac{m}{m-1} = \frac{2m-6}{2m}$$

(A) 1/2 (B) 5/8 (C) 7/8 (D) 3/4 (E) 2/3

Cross-multiplying yields

$$m \cdot 2m = (m-1)(2m-6)$$
$$2m^2 = 2m^2 - 6m - 2m + 6$$
$$-6 = -8m$$
$$m = 3/4$$

Answer: (D) 3/4

54. Given a and b are integers, $-5 < a < -4$ and $0 < b < 3$, what is the largest approximate value for $a^2 + b^2$? Round off to the nearest integer.

(A) 36 (B) 34 (C) 37 (D) 32 (E) 33

The largest $a^2 = (-4.999...)^2 \approx 24.999...$
+ The largest $b^2 = (2.999...)^2 \approx 8.999...$
$\phantom{+ \text{The largest } b^2 = (2.999...)^2 \approx\ } 33.999...$

Answer: (B) 34

Method II:

Squaring the terms of the inequality $0 < b < 3$ will preserve the direction of the inequality because all the terms are positive:

$$0 < b^2 < 9$$

Squaring the terms of the inequality $-5 < a < -4$ will reverse the direction of the inequality because all the terms are negative:

$$25 > a^2 > 16$$

Rewriting this inequality in standard form, with the smaller number on the left and the larger number on the right, gives

$$16 < a^2 < 25$$

So, we have the following system of two inequalities:

$$16 < a^2 < 25$$
$$0 < b^2 < 9$$

Adding the terms of these inequalities gives

$$16 + 0 < a^2 + b^2 < 25 + 9$$
$$16 < a^2 + b^2 < 34$$

55. Given that x and y are positive integers and $x > y$, which of the following choices is the best description of $\dfrac{|x-y|}{|x+y|}$.

(A) 1   (B) < 1   (C) > 1   (D) 2   (E) > 2

Since $x > y$, $x - y > 0$. Now, clearly, the sum of two positive numbers is greater than their difference:

$$0 < x - y < x + y$$

Dividing this inequality by $x + y$ (a positive number) yields

$$\frac{0}{x+y} < \frac{x-y}{x+y} < \frac{x+y}{x+y}$$

$$0 < \frac{x-y}{x+y} < 1$$

Taking the absolute value of this inequality yields

$$\left|\frac{x-y}{x+y}\right| = \frac{|x-y|}{|x+y|} < 1$$

Answer: (B) < 1

56. Water runs into a pool 9 meters long by 5 meters wide and 4 meters high at the rate of 120,000 cubic centimeters per minute. How long will it take to fill the pool?
l meter = 100 centimeters

(A) 25 hours   (B) 46 hours   (C) 30 hours   (D) 50 hours   (E) 38 hours

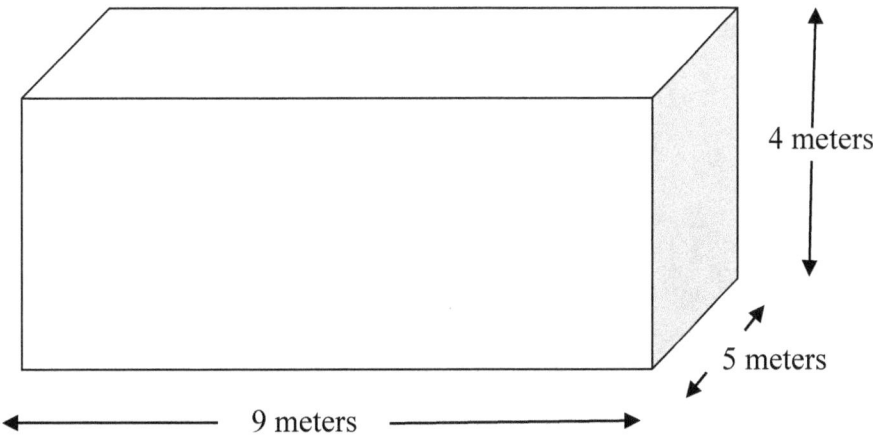

9 meters = 900 centimeters
5 meters = 500 centimeters
4 meters = 400 centimeters

$$\frac{Volume\ of\ the\ pool}{Rate\ being\ filled} = \frac{900 \cdot 500 \cdot 400\ cubic\ centimeters}{120,000\ cubic\ centimeters/minute} = 1,500\ minutes$$

Converting 1,500 minutes into hours gives

$$\frac{1,500\ minutes}{60\ minutes\ in\ a\ hour} = 25\ hours$$

Answer: (A) 25 hours

57. The gas tank of a car holds 25 gallons. The car uses one gallon per 22 miles driving on a horizontal road. It uses one gallon per 18 miles driving uphill. The driver has gone 462 miles on a horizontal road. For the rest of trip, it's uphill driving. How many more miles will the remainder of gas in the tank support driving on the uphill road?

(A) 84 (B) 72 (C) 104 (D) 98 (E) 85

horizontal road:

$$\frac{Distance\ traveled}{Miles\ per\ gallon} = \frac{462}{22} = 21\ gallons\ used$$

So, there are 4 (= 25 − 21) gallons remaining for the uphill drive.

uphill driving:

$$4\ gallons \times 18\ mpg = 72\ miles$$

Answer: (B) 72

58. Find the greatest prime number, x, that will satisfy the inequality.

$$2 \leq \sqrt{\frac{x+6}{3}} \leq \sqrt{47}$$

(A) 413 (B) 133 (C) 247 (D) 323 (E) 477

Since all the expressions are greater than 0, squaring them will not change the direction of the inequalities:

$$(2)^2 \leq \left(\sqrt{\frac{x+6}{3}}\right)^2 \leq (\sqrt{47})^2$$

$$4 \leq \frac{x+6}{3} \leq 47$$

Multiplying by 3 yields

$$12 \leq x+6 \leq 141$$

$$6 \leq x \leq 135$$

Greatest prime number x to satisfy the inequality is 133.

Answer: (B) 133

59. Select the set of equations to describe the area common to both in the figure shown.

(A) $y = 3 \cap y < x + 2$
(B) $y \geq 3 \cap y < x + 2$
(C) $y \leq 3 \cap y = x + 2$
(D) $y \leq x + 2 \cap y < 3$
(E) $y < 3 \cap y \leq x + 2$

We are looking for the region that contains both the vertical dashed lines and the diagonal dashed lines:

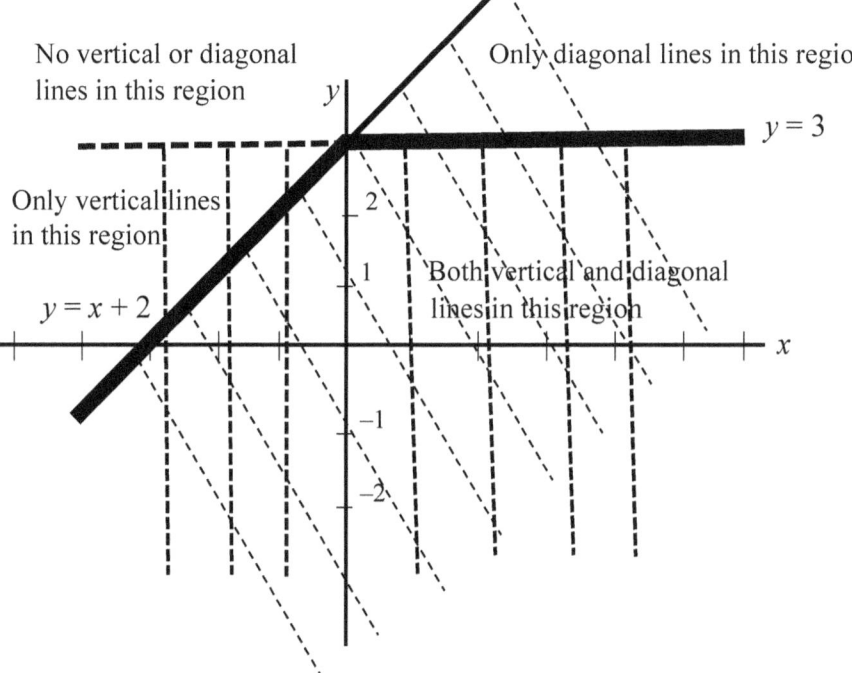

Answer: (D) $y \leq x + 2 \cap y < 3$

60. A metal pitcher filled with water weighs 21 ounces. If it's half full, the pitcher and the water weigh 12 ounces. How many ounces does the pitcher weigh alone?

(A) 5 ounces   (B) 3 ounces   (C) 6 ounces   (D) 4 ounces   (E) 2 ounces

Let x = the weight of the pitcher.
Let y = the weight of the water in the half-full pitcher.
Let $2y$ = the weight of the water in the full pitcher.

Total weight in the half-full pitcher: (1) $x + y = 12$
Total weight in the full pitcher: (2) $x + 2y = 21$

                                                                    (1) $x = 12 - y$
Substitute $12 - y$ for x in (2):   (2) $(12 - y) + 2y = 21$
                                                                    (2) $12 + y = 21$
Weight of the water in the half-full pitcher: (2) $y = 9$

Go back to (1) and substitute 9 for y: (1) $x + 9 = 12$
                                                                    (1) $x = 3$

Answer: (B) 3 ounces

Answer Key to Test 4

1. D	21. B	41. B
2. C	22. D	42. E
3. B	23. A	43. B
4. E	24. B	44. C
5. D	25. A	45. E
6. C	26. A	46. A
7. D	27. B	47. D
8. A	28. E	48. C
9. E	29. C	49. D
10. C	30. D	50. E
11. D	31. D	51. A
12. D	32. D	52. A
13. C	33. A	53. D
14. E	34. A	54. B
15. D	35. B	55. B
16. A	36. C	56. A
17. B	37. A	57. B
18. E	38. E	58. B
19. B	39. C	59. D
20. C	40. A	60. B

Test 5

1. Which of the following choices for x <u>does not</u> satisfy the inequality $2\sqrt{2x-3} \geq 4$?

    (A) 4   (B) 3.9   (C) 3.5   (D) 3.2   (E) 5.3

2. Let $a = x^{2/3}$ and let $x = b^6$. Find the value of $(ab)^{2/5}$ in terms of x.

    (A) $x^{1/3}$ (B) $x^{1/6}$ (C) $x^{2/3}$ (D) $x^{3/8}$ (E) $x^{1/9}$

3. If x and y represent positive integers, find the median of $y-2$, x, $x+6$, $y+4$, $y-3$, and $x-3$ if $y = x+4$.

    (A) $2x + .5$ (B) $x + 1.5$ (C) $2x - .5$ (D) $\dfrac{2x+4}{2}$ (E) $x - 3$

4. Matrix A below indicates the average number of foreign tourists who frequent restaurants Alpha and Beta each day. Matrix B indicates the average price the tourists paid for dinner. What is the average total amount of money the tourists paid for dinner each day?

	A				B
	British	Japanese	German		Cost per Meal ($)
alpha	50	60	40	British	40
beta	60	40	70	Japanese	50
				German	30

    (A) $12,700   (B) $9,800   (C) $10,500   (E) $11,600   (F) $10,200

5. In the general quadratic equation $ax^2 + bx + c = 0$, if $a = 1$, $b = -4$, and one root is 6, what is a possible value for c?

    (A) 3   (B) -12 (C) 8   (D) -6 (E) 9

6. Find the greatest common factor of $3x^2y^3z^4$, $9xy^2z^3$ and $6x^2yz^2$.

    (A) $9xy^3z^2$ (B) $3xyz^2$ (C) $12x^2y^2z^2$ (D) $3xy^3z^3$ (E) $3x^2y^4z^3$

7. If the point $(g, 4g)$ lies on a line whose slope is 2, what is the y-intercept?

(A) $3g$ (B) $4g$ (C) $2g$ (D) g (E) $g + 2$

8. List the <u>maximum</u> number of operations under which the set of integers is closed.

(a) addition (b) subtraction (c) multiplication (d) division

(A) a, b, d (B) b, c, d (C) a, d (D) a, b, c (E) a, b, c, d

9. Let the definition of $x \# y = \dfrac{x-y}{2}$ and let the definition of $x * y = x^2 + y$ for any real numbers x and y. Find the value of $(2 \# 3) * (4 * 3)$.

(A) $12\dfrac{3}{4}$ (B) $18\dfrac{3}{4}$ (C) $16\dfrac{3}{4}$ (D) $18\dfrac{1}{2}$ (E) $19\dfrac{1}{4}$

10. Find the fourth vertex of the parallelogram with vertices A(7, 5), B(2, 1) and C(10, 1).

(A) (15, 5) (B) (12, 5) (C) (10, 3) (D) (10, 5) (E) (15, 3)

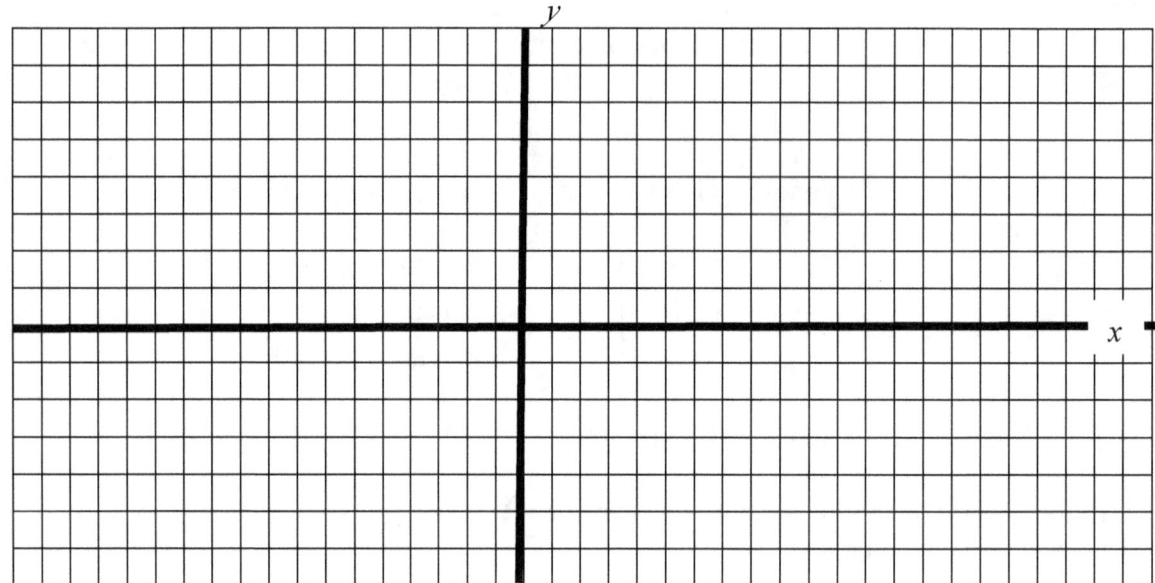

11. Given $\frac{2}{9}|b-2|^2 \geq 18$. Find a possible value for b.

(A) –6 (B) 10 (C) 8 (D) –5 (E) –7

12. In a geometric sequence the first term is 3. If the second term is $3/a$, find the 6th term.

(A) $\frac{3}{a^5}$ (B) $9a^5$ (C) $\frac{243}{a^5}$ (D) $243a^5$ (E) $3a^5$

13. An airplane pilot is flying from city A to city B. The distance between the two cities is d miles. If she's already flown a fraction, b/c, of the way, how many miles does she still have to fly?

(A) $\frac{bd}{c}$ (B) $c - \frac{bd}{c}$ (C) $d - \frac{bd}{c}$ (D) $\frac{bd-d}{c}$ (E) $\frac{b-bd}{c}$

14. Out of a total of b computer chips, c pass inspection. What fraction fail?

(A) $\frac{b-c}{b}$ (B) $\frac{c-b}{c}$ (C) $\frac{c-b}{b}$ (D) $\frac{b+c}{c}$ (E) $\frac{b-c}{c}$

15. Name one value for x that makes this fraction undefined.

$$\frac{2x^2 + 3x - 4}{x^2 - 7x + 12}$$

(A) 6 (B) 3 (C) 5 (D) 2 (E) 1

16. A cylinder 7 feet high has a volume 31 cubic feet greater than a cube whose edge is 7 feet. Find the radius of the cylinder. Let $\pi = 22/7$.

(A) $3\sqrt{7}$ (B) $2\sqrt{5}$ (C) $\sqrt{19}$ (D) $\sqrt{17}$ (E) $3\sqrt{15}$

17. If we multiply $\dfrac{3b^2-4}{2}$ by 3, we get 282. What is the negative value of b?

(A) –10   (B) –7   (C) –6   (D) –8   (E) –9

18. In a coffee harvest, 20 ounces of coffee are rejected out of every 90 pounds. What is the ratio of non-rejected coffee to rejected coffee? One pound = sixteen ounces.

(A) 19 : 3   (B) 56 : 1   (C) 26 : 1   (D) 71 : 1   (E) 32 : 3

19. What is the range of the function $f(x) = 3 \cos 5x$?

(A) $-5 \le y \le 5$ (B) $-3 < y < 3$ (C) $-5 < y < 5$ (D) $-3 \le y \le 3$ (E) $-3.5 \le y \le 3.5$

20. Carmen and Mandy run towards each other from towns A and B, respectively. The towns are 40 miles apart. Carmen runs at the rate of 8 mph. Mandy starts 1/2 hour later and runs 2 mph faster than Carmen. If they meet at 3:15 pm, at what time did Carmen start running?

(A) 12:45 pm   (B) 11:30 am   (C) 1:15 pm   (D) 1 pm   (E) 12:15 pm

21. A parabola is the set of points equidistant from a fixed line, called the directrix, and a fixed point, called the focus. Using this definition, develop an equation for a parabola with a focus of (2, 0) and a directrix $x = -2$, by setting the two distances equal.

(A) $y = \pm\sqrt{2x+4}$ (B) $x = 6y^2$ (C) $y = \pm\sqrt{3x}$ (D) $y = 6x^2$ (E) $y = \pm\sqrt{8x}$

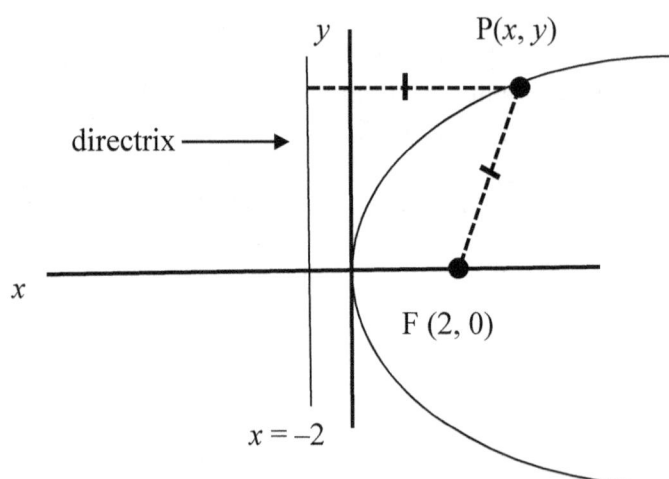

22. If the central angle of a circle measures 60° and the radius is 6, find the shaded area (the segment). Leave your answer in terms of π.

(A) $4\pi - 3\sqrt{2}$ (B) $6\pi - 8\sqrt{3}$ (C) $12\sqrt{3} - 4\pi$ (D) $6\sqrt{3} - 4\pi$ (E) $6\pi - 9\sqrt{3}$

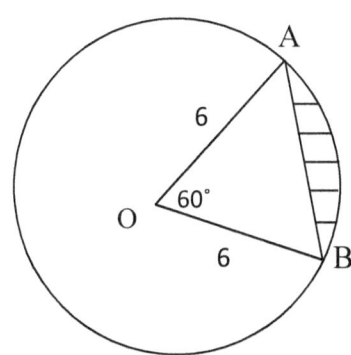

23. The letters a, b and c represent 3 positive integers. If we double the value of a, reduce the value of b by $\frac{3}{5}b$ and multiply the value of c by $2\frac{1}{2}$, compare the value of the product of a, b and c with the product of the new values.

(A) 1 : 2 (B) 1 : 3 (C) 2 : 3 (D) 2 : 1 (E) 2 : 5

24. A group of weightlifters competed. The results, by lifted weight, are listed below. Find the mean, correct to the nearest tenth.

(A) 149.1 (B) 157.4 (C) 158.6 (D) 166.5 (E) 161.2

Pounds	Number of Weightlifters
124	2
153	4
167	3
175	2
192	1

25. Let $f(x, y) = 3x^2 - 2y - 68 \geq 0$. If $y = 20$, what is the maximum negative value that x can assume?

(A) 4 (B) –5 (C) –6 (D) 7 (E) 8

26. In the quadrilateral shown AB = CD and AB ∥ CD. List the maximum number of true statements.

(1) BC ≅ AD   (2) AB ≅ BC   (3) AE + BE = EC + ED   (4) AE ≅ BE   (5) BC ∥ AD

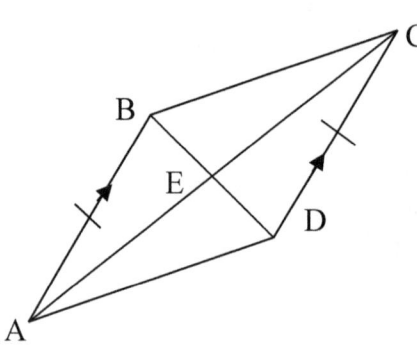

(A) 1, 2   (B) 2, 4   (C) 1, 2, 4   (D) 1, 3, 5   (E) 2, 4, 5

27. In circle O, radius AO = 3 and $m\angle B = 60°$. Find the length of the minor arc $\overarc{AB}$. Leave your answer in terms of π.

(A) π (B) $\pi/3$ (C) $2\pi/3$ (D) 2π (E) $3\pi/4$

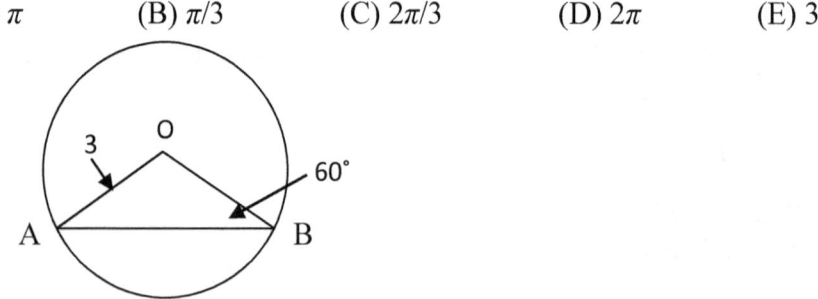

28. Find the sum of the first 12 terms of the geometric sequence 2, 4, 8, 16,....

(A) $-2(1-2^{13})$ (B) $2(1-2^{13})$ (C) $-2(1-2^{13})$ (D) $-2(1-2^{12})$ (E) $-(1-2^{12})$

29. A and B are two independent events out of many. If the probability of A occurring is .2 and the probability of B occurring is .6, what is the probability of A and not B?

(A) .12   (B) .24   (C) .08   (D) .32   (E) .48

30. Find the inverse of the logarithmic function $f(x) = \log_3(x + 2)$.

(A) $2^x - 3$ (B) $3^2 - x$ (C) $2^x + 3$ (D) $2 - 3^x$ (E) $3^x - 2$

31. The letters *a*, *b* and *c* are positive consecutive even integers and the letters *d*, *e*, and *f* are positive consecutive odd integers such that *d* is 3 more than twice *b*. Describe *f* in terms of *a*.

(A) $3a - 3$ (B) $2a + 6$ (C) $2a + 3$ (D) $2b + 7$ (E) $2a + 11$

32. The Fahrenheit temperature for boiling water is 212°. The Celsius temperature for boiling water is 100°. The Fahrenheit temperature for freezing water is 32°, while the corresponding Celsius temperature for freezing water is 0°. The formula linking the two scales together is

$$F = \frac{9}{5}C + 32$$

At what temperature do the scales have the same value?

(A) 40° (B) –40° (C) 60° (D) –60° (E) –50°

33. A train traveling at 80 mph enters a tunnel. If it takes 6 seconds from the time the train enters the tunnel until the first car reaches the far end of the tunnel, how long is the train? One mile = 5,280 feet.

(A) 648 ft. (B) 586 ft. (C) 680 ft. (D) 704 ft. (E) 726 ft.

34. Find the radius of the circle shown if the area of the sector is 154/8 and the central angle is 45°. Let $\pi = 22/7$.

(A) 3 (B) 4 (C) 5 (D) 6 (E) 7

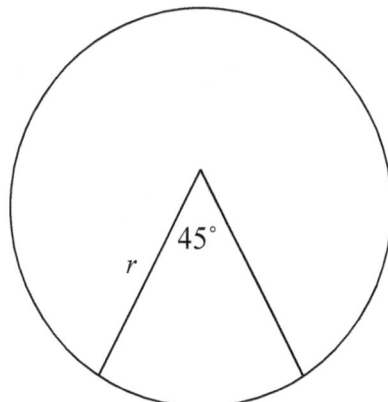

35. Margaret is three times as old as her daughter. Four years ago Margaret was four times as old as her daughter was then. How old is Margaret now?

(A) 45 (B) 39 (C) 36 (D) 28 (E) 24

36. Which equation represents a move up by 3 units of the original equation $y = (x + 2)^2$?

(A) $y = x^2 + 3$ (B) $y = 3(x - 2)^2 + 2$ (C) $y = x^2 + 4x + 7$ (D) $y = x^2 + 2$
(E) $y = 2x^2 + 3$

37. If each interior angle of a regular polygon is 20° more than 3 times its adjacent exterior angle, how many sides does the polygon have? Note: The formula linking the number of degrees (a) in an individual interior angle of a regular polygon to the number of sides (n) is

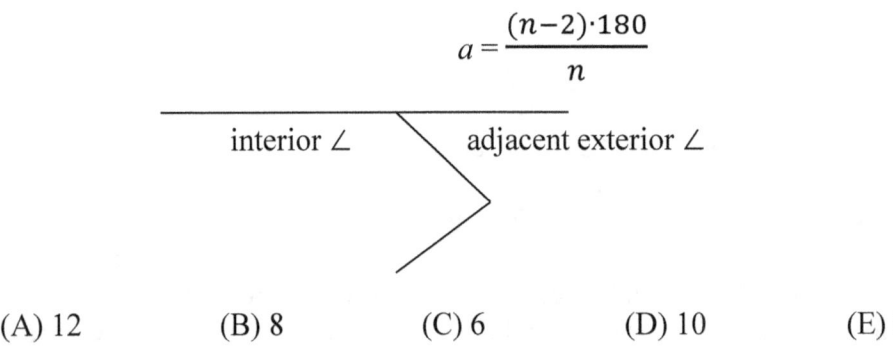

(A) 12 (B) 8 (C) 6 (D) 10 (E) 9

38. Jamal dives into a river and swims across at an angle of 45° with the terrain. When he touches land he has completed a swim of 2,400 feet. How wide is the river? Round your answer to the nearest foot.

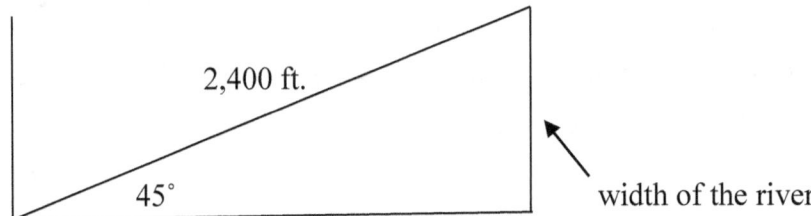

(A) 2,435 ft. (B) 1,697 ft. (C) 1,982 ft. (D) 2,072 ft. (E) 2,123 ft.

39. Jack jogs 12 miles at the rate of 8 mph. He walks back along the same path at the rate of 3 mph. What was his average speed?

(A) 5.2 mph   (B) 4.5 mph   (C) 4.6 mph   (D) 5.5 mph   (E) 4.4 mph

40. If $(x - y)(x - y) = 81$ and $y = -3$, find one value of x.

(A) 5   (B) –12   (C) –5   (D) 3   (E) –4

41. Select the number line which expresses the range of values for x in the absolute inequality $|3x - 2| < 10$.

(A) ![number line with open circles at -2⅔ and 4, segment between]   (B) ![number line with closed circle at -2⅔ and open circle at 4, segment between]   (C) ![number line with closed circles at -2⅔ and 4, segment between]

(D) ![number line with open circle at -2⅔ and closed circle at 4, segment between]   (E) ![number line with open circle at -2⅔ and closed circle at 4, segment outside]

42. Find the reflection of point P (4, 2) across the line $y = x$.

(A) (2, 4)   (B) (4, 3)   (C) (3, 5)   (D) (–3, 4)   (E) (4, –3)

43. Find the y-coordinate of point A on the ellipse shown, with center at the origin. Leave your answer in radical form.

(A) $\sqrt{5.25}$ (B) $\sqrt{6.75}$ (C) $\sqrt{3.85}$ (D) $\sqrt{4.15}$ (E) $2\sqrt{4.85}$

$$\frac{x^2}{a^2} + \frac{y^2}{b^2} = 1$$

where a is 1/2 the length of the horizontal axis and b is 1/2 the length of the vertical distance axis x and y are coordinates of any one point along the ellipse.

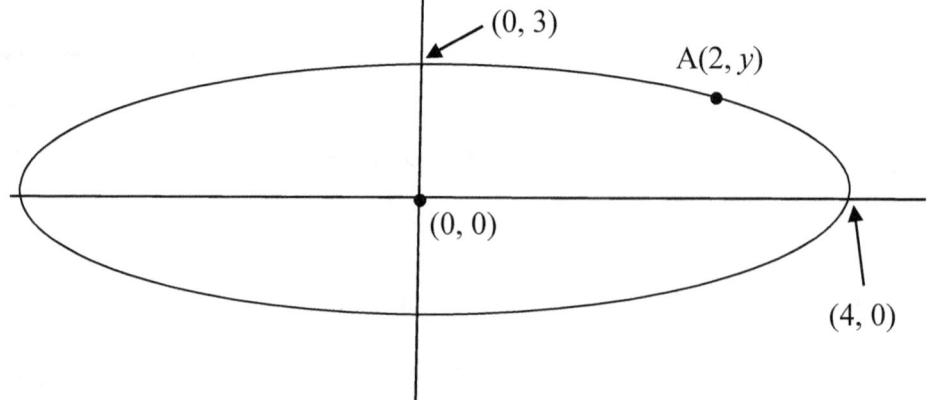

44. If $\frac{3}{5}$ of 3% of $b = 81$, find $\frac{2b}{3}$.

(A) 6,000 (B) 4,500 (C) 6,500 (D) 3,000 (E) 2,500

45. Solve for the two values of x in the equation $\sqrt{10x+1} = \sqrt{9x-8} + 1$.

(A) $x = 5, 6$ (B) $x = 7, 4$ (C) $x = 3, 8$ (D) $x = 9, 11$ (E) $x = 8, 12$

46. Name the point in the first quadrant where the graphs of the functions $f(x) = 6/x$ and $g(x) = x$ cross?

(A) $(\sqrt{3}, \sqrt{3})$ (B) $(2\sqrt{3}, -2\sqrt{3})$ (C) $(\sqrt{6}, \sqrt{6})$ (D) $(\sqrt{5}, -\sqrt{5})$ (E) $(3\sqrt{3}, -3\sqrt{3})$

47. Five pounds of Colombian coffee at $4.90 per pound are mixed with eight pounds of Venezuelan coffee to make a mixture at $5.70 per pound. What is the price of the Venezuelan coffee?

(A) $6.20   (B) $5.40   (C) $5.10   (D) $5.30   (E) $6.30

48. In right triangle ABC, DE ∥ AB, AC = 10, AB = 15, DE is 9 more than BE, and DC = 8. Find BE.

(A) 3   (B) 5   (C) 9   (D) 12   (E) 8

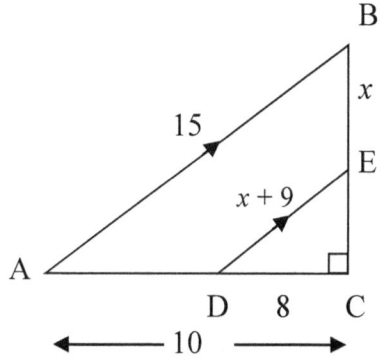

49. What is the width of the parabola representing the equation $y^2 = 12x$ at $x = 3$?

(A) 8   (B) 6   (C) 10   (D) 12   (E) 14

50. If $\frac{1}{r} = \frac{1}{s} - \frac{1}{t}$, find the value of r in terms of s and t.

(A) $\frac{s+t}{t-s}$ (B) $\frac{st}{t-s}$ (C) $\frac{t-s}{t+s}$ (D) $\frac{s+t}{s-t}$ (E) $\frac{ts}{s+t}$

51. We have 8 different colored paints. We'll call them a, b, c, d, e, f, g and h. We want to paint 3 rooms. However, we don't want to use paint b in rooms 1 or 2. Once a paint is used, we cannot use it again in another room. In how many different ways can we paint the three rooms, using only one color per room?

(A) 336   (B) 384   (C) 480   (D) 252   (E) 216

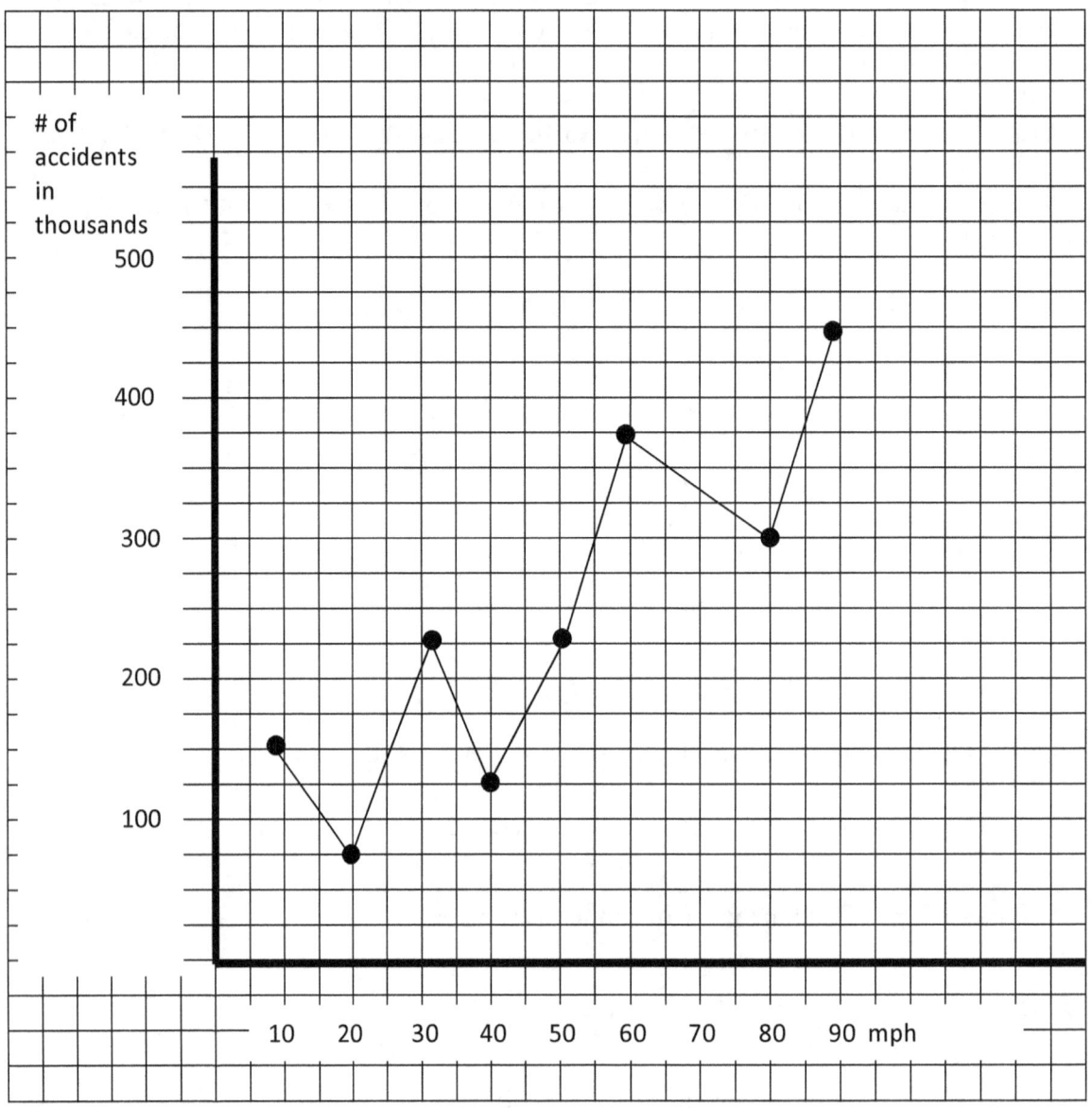

52. What was the percent increase in the number of accidents in driving at 40 mph compared to 80 mph? Round off to the nearest percent.

(A) 90% (B) 120% (C) 140% (D) 180% (E) 330%

53. If the volume of this triangular prism is 96 and the base of the triangle is 6 and the height of the triangle is 4, find the length of the prism.

(A) 6　　　　(B) 10　　　　(C) 8　　　　(D) 12　　　　(E) 9

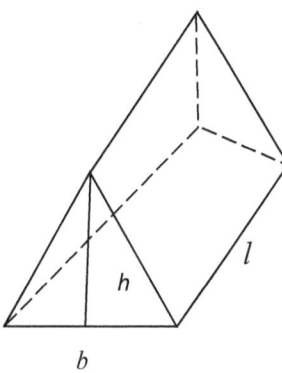

54. Find the value of y in the trigonometric equation $y = 4 \cos x + 1$ when $x = 2\pi$.

(A) 5　　　　(B) 2　　　　(C) 3　　　　(D) 9　　　　(E) 8

55. The center of a circle is located on the xy-plane at $(-1, 2)$. Its radius is $\sqrt{29}$. At which positive point does the circle cross the x-axis?

(A) (3, 0)　　　(B) (0, 3)　　　(C) (4, 0)　　　(D) (5, 0)　　　(E) (0, 4)

56. Find the area of the triangle enclosed by the three squares illustrated below. Leave your answer in radical form.

(A) 20　　　　(B) 8　　　　(C) 18　　　　(D) 12　　　　(E) 6

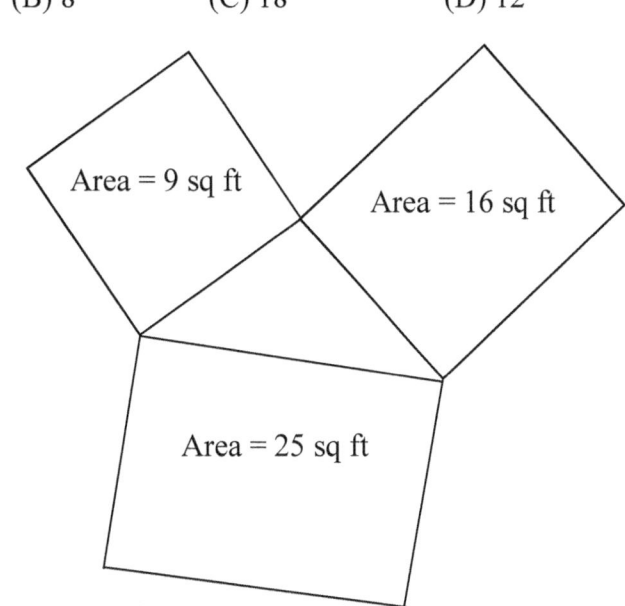

57. There are 4 green balls, 6 red balls and 3 yellow balls in a container. What is the probability of first selecting a red ball, not replacing it, and then selecting a yellow ball?

(A) 7/19   (B) 3/26   (C) 8/27   (D) 5/16   (E) 4/9

58. If one point on the graph of the function $y = (x - 2)^3$ is A(4, y_1), what is the equation of the straight line parallel to the x-axis which passes through that point?

(A) $y = 8$ (B) $x = 8$ (C) $y = 4x$ (D) $x = 6y$ (E) $y = -4$

59. Which of the lines below has a slope of –2?

(A)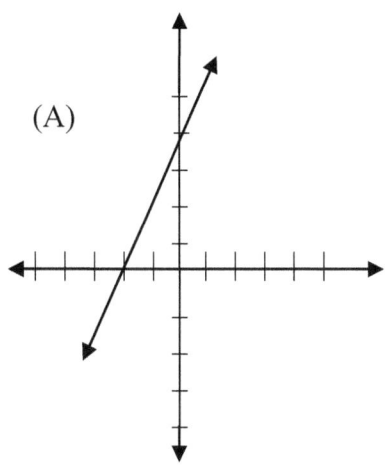

(B)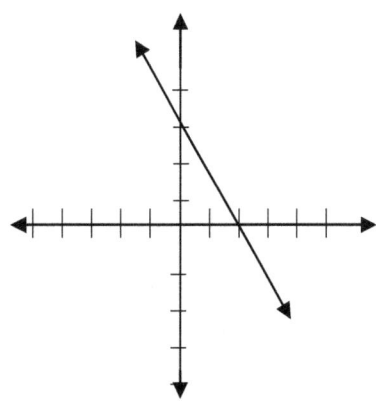

(C)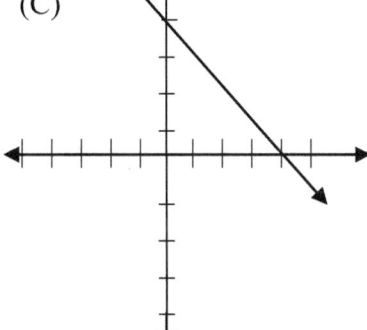

(D)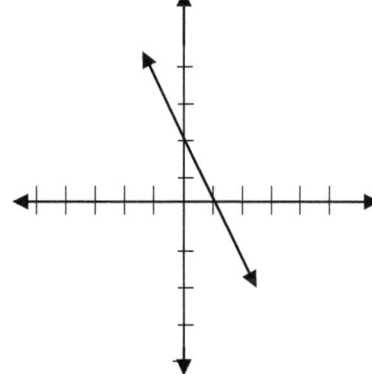

(E)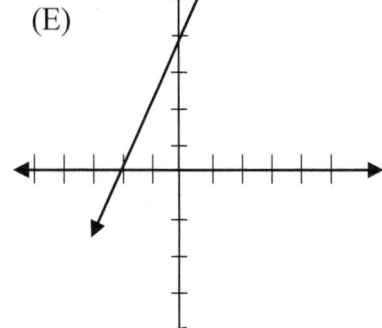

60. Find the average rate of change of the function displayed from $x = -4$ to $x = 3$.

(A) 2/5   (B) 5/8   (C) 3/7   (D) 4/9   (E) 5/7

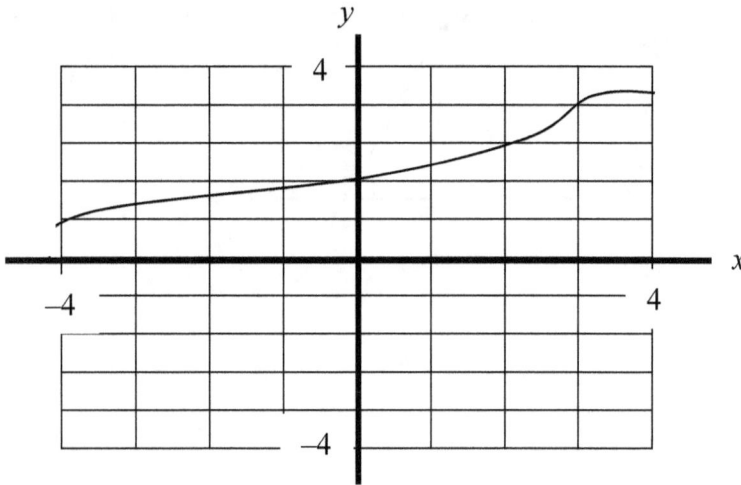

Answers to Test 5

1. Which of the following choices for x <u>does not</u> satisfy the inequality $2\sqrt{2x-3} \geq 4$?

(A) 4   (B) 3.9   (C) 3.5   (D) 3.2   (E) 5.3

Square both sides of the inequality, which does not change the direction of the inequality since both sides are greater than 0:

$$4(2x-3) \geq 16$$

$$8x - 12 \geq 16$$

$$8x \geq 28$$

$$x \geq 3.5$$

Now, 3.2 is the only answer-choice offered that is not larger than 3.5.

Answer: (D) 3.2

2. Let $a = x^{2/3}$ and let $x = b^6$. Find the value of $(ab)^{2/5}$ in terms of x.

(A) $x^{1/3}$ (B) $x^{1/6}$ (C) $x^{2/3}$ (D) $x^{3/8}$ (E) $x^{1/9}$

$a = x^{2/3}$
$x = b^6, b = x^{1/6}$

$$(ab)^{2/5}$$
$$(x^{2/3} \cdot x^{1/6})^{2/5}$$
$$(x^{5/6})^{2/5}$$
$$x^{10/30}$$
$$x^{1/3}$$

Answer: (A) $x^{1/3}$

3. If x and y represent positive integers, find the median of $y - 2$, x, $x + 6$, $y + 4$, $y - 3$, and $x - 3$ if $y = x + 4$.

(A) $2x + .5$ (B) $x + 1.5$ (C) $2x - .5$ (D) $\dfrac{2x+4}{2}$ (E) $x - 3$

Original terms: $y - 2$, x, $x + 6$, $y + 4$, $y - 3$, $x - 3$
$y = x + 4$: $(x + 4) - 2$, x, $x + 6$, $(x + 4) + 4$, $(x + 4) - 3$, $x - 3$
Adding: $x + 2$, x, $x + 6$, $x + 8$, $x + 1$, $x - 3$

Arranging in order, from smallest to largest, yields

$$x - 3, \quad x, \quad x + 1, \quad x + 2, \quad x + 6, \quad x + 8$$

$$\text{Median} = \frac{(x + 1) + (x + 2)}{2} = \frac{2x + 3}{2} = x + 1.5$$

Answer: (B) $x + 1.5$

4. Matrix A below indicates the average number of foreign tourists who frequent restaurants Alpha and Beta each day. Matrix B indicates the average price the tourists paid for dinner. What is the average total amount of money the tourists paid for dinner each day?

$$
A = \begin{array}{c} \\ \text{alpha} \\ \text{beta} \end{array}
\begin{array}{ccc} \text{British} & \text{Japanese} & \text{German} \\ \left[\begin{array}{ccc} 50 & 60 & 40 \\ 60 & 40 & 70 \end{array}\right] \end{array}
\qquad
B = \begin{array}{c} \text{Cost per Meal (\$)} \\ \begin{array}{c} \text{British} \\ \text{Japanese} \\ \text{German} \end{array} \left[\begin{array}{c} 40 \\ 50 \\ 30 \end{array}\right] \end{array}
$$

(A) $12,700 (B) $9,800 (C) $10,500 (E) $11,600 (F) $10,200

Multiply across rows and down columns and add the results:

(First row times column) + (Second row times column)

$$(50 \cdot 40 + 60 \cdot 50 + 40 \cdot 30) + (60 \cdot 40 + 40 \cdot 50 + 70 \cdot 30) =$$

$$12{,}700$$

Answer: (A) $12,700

5. In the general quadratic equation $ax^2 + bx + c = 0$, if $a = 1$, $b = -4$, and one root is 6, what is a possible value for c?

(A) 3   (B) –12   (C) 8   (D) –6   (E) 9

Plug in the information:

$$ax^2 + bx + c = 0$$

$a = 1, b = -4$:

$$x^2 - 4x + c = 0$$

Now, look at choice (B):

$$x^2 - 4x - 12 = 0$$

$$(x + 2)(x - 6) = 0$$

Answer: (B) –12

6. Find the greatest common factor of $3x^2y^3z^4$, $9xy^2z^3$ and $6x^2yz^2$.

(A) $9xy^3z^2$ (B) $3xyz^2$ (C) $12x^2y^2z^2$ (D) $3xy^3z^3$ (E) $3x^2y^4z^3$

Look for the greatest common digit. Then look for the greatest common variable.

    Greatest common factors: 3, x, y and z^2
    The single greatest common factor is $3xyz^2$

Answer: (B) $3xyz^2$

7. If the point $(g, 4g)$ lies on a line whose slope is 2, what is the y-intercept?

(A) $3g$ (B) $4g$ (C) $2g$ (D) g (E) $g + 2$

$y = mx + b$, where m = slope and b = the y-intercept

$m = 2, (g, 4g)$: $4g = 2g + b$
                       $b = 2g$

Answer: (C) $2g$

8. List the <u>maximum</u> number of operations under which the set of integers is closed.

(a) addition   (b) subtraction   (c) multiplication   (d) division

(A) a, b, d   (B) b, c, d   (C) a, d   (D) a, b, c   (E) a, b, c, d

We can add, subtract and multiply integers and get an integer answer. However, we cannot divide all integers and get an integer answer (example: 5/2). Therefore, the set of integers is closed for the operations of addition, subtraction and multiplication only.

Answer: (D) a, b, c

9. Let the definition of $x \# y = \frac{x-y}{2}$, and let the definition of $x * y = x^2 + y$ for any real numbers x and y. Find the value of $(2 \# 3) * (4 * 3)$.

(A) $12\frac{3}{4}$ (B) $18\frac{3}{4}$ (C) $16\frac{3}{4}$ (D) $18\frac{1}{2}$ (E) $19\frac{1}{4}$

$$x \# y = \frac{x - y}{2} \qquad\qquad x * y = x^2 + y$$

$$2 \# 3 = \frac{2 - 3}{2} \qquad\qquad 4 * 3 = 4^2 + 3$$

$$2 \# 3 = \frac{-1}{2} \qquad\qquad 4 * 3 = 19$$

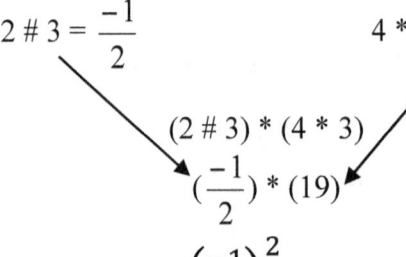

$$(2 \# 3) * (4 * 3)$$
$$\left(\frac{-1}{2}\right) * (19)$$
$$\left(\frac{-1}{2}\right)^2 + 19$$
$$\frac{1}{4} + 19$$
$$19\frac{1}{4}$$

Answer: (E) $19\frac{1}{4}$

10. Find the fourth vertex of the parallelogram with vertices A(7, 5), B(2, 1) and C(10, 1).

(A) (15, 5)   (B) (12, 5)   (C) (10, 3)   (D) (10, 5)   (E) (15, 3)

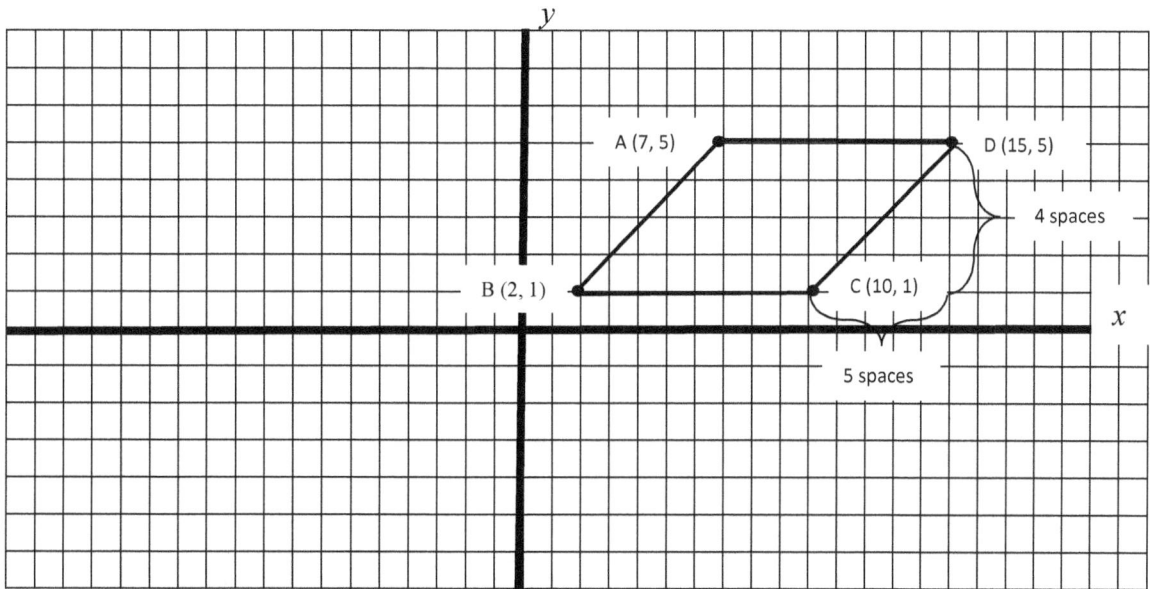

From Point C, we count 5 horizontal spaces, which takes us from 10 to 15. Then from the point (15, 1), we count 4 vertical spaces, which takes us from 1 to 5. So, the fourth vertex is (15, 5).

Answer: (A) (15, 5)

ACT Math Personal Tutor

11. Given $\frac{2}{9}|b-2|^2 \geq 18$. Find a possible value for b.

(A) –6   (B) 10   (C) 8   (D) –5   (E) –7

Try $b = -7$:

$$\frac{2}{9}|-7-2|^2 \geq 18$$

$$\frac{2}{9}|-9|^2 \geq 18$$

$$\frac{2}{9}(81) \geq 18$$

$$18 \geq 18 \checkmark$$

Answer: (E) $b = -7$

12. In a geometric sequence the first term is 3. If the second term is $3/a$, find the 6th term.

(A) $\frac{3}{a^5}$ (B) $9a^5$ (C) $\frac{243}{a^5}$ (D) $243a^5$ (E) $3a^5$

If the first term is 3, we have to multiply by $\frac{1}{a}$ to get the second term ($3/a$) and therefore the common ratio is $\frac{1}{a}$. For the 6th term, let's use the formula $a_n = a_1 r^{n-1}$, where a_n = the nth term, a_1 = the first term and r = the common ratio.

$$a_6 = 3\left(\frac{1}{a}\right)^{6-1}$$

$$= 3\left(\frac{1}{a}\right)^5$$

$$= \frac{3}{a^5}$$

Answer: (A) $\frac{3}{a^5}$

13. An airplane pilot is flying from city A to city B. The distance between the two cities is d miles. If she's already flown a fraction, b/c, of the way, how many miles does she still have to fly?

(A) $\dfrac{bd}{c}$ (B) $c - \dfrac{bd}{c}$ (C) $d - \dfrac{bd}{c}$ (D) $\dfrac{bd-d}{c}$ (E) $\dfrac{b-bd}{c}$

$\dfrac{b}{c} \cdot d$ (fraction already flown • total = number of miles already flown)

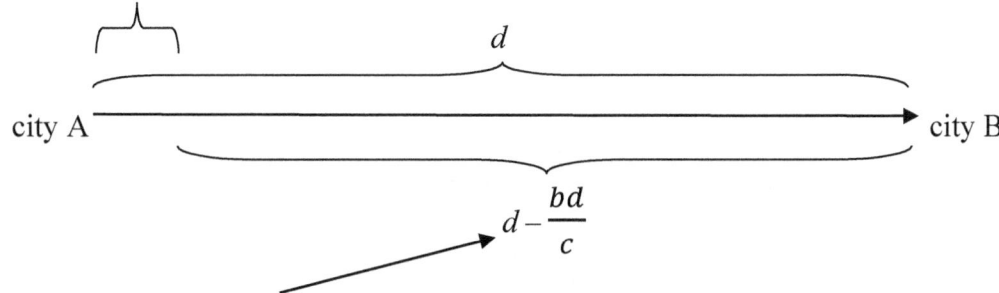

She still has to fly $d - \dfrac{bd}{c}$ miles to go.

Answer: (C) $d - \dfrac{bd}{c}$

ACT Math Personal Tutor

14. Out of a total of b computer chips, c pass inspection. What fraction fail?

(A) $\dfrac{b-c}{b}$ (B) $\dfrac{c-b}{c}$ (C) $\dfrac{c-b}{b}$ (D) $\dfrac{b+c}{c}$ (E) $\dfrac{b-c}{c}$

If c out of b computer chips pass, then $b - c$ fail. The total number of computer chips is b.

$$\frac{Fail}{Total} = \frac{b-c}{b}$$

Answer: (A) $\dfrac{b-c}{b}$

15. Name <u>one</u> value for x that makes this fraction undefined.

$$\frac{2x^2 + 3x - 4}{x^2 - 7x + 12}$$

(A) 6 (B) 3 (C) 5 (D) 2 (E) 1

If the denominator is zero, the entire fraction is undefined (cannot divide by zero). So set the denominator equal to zero, factor it and determine which values of x makes the denominator zero.

$$x^2 - 7x + 12 = 0$$

$$(x-4)(x-3) = 0$$

$$x = 4 \text{ or } x = 3$$

So, either value for x, 4 or 3, makes the denominator zero.

Answer: (B) 3

16. A cylinder 7 feet high has a volume 31 cubic feet greater than a cube whose edge is 7 feet. Find the radius of the cylinder. Let $\pi = 22/7$.

(A) $3\sqrt{7}$ (B) $2\sqrt{5}$ (C) $\sqrt{19}$ (D) $\sqrt{17}$ (E) $3\sqrt{15}$

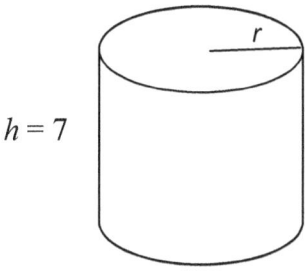

Volumes:

$V_c = \pi r^2 h$ $\qquad\qquad V_b = e^3$

$V_c = \pi r^2 (7)$ $\qquad\qquad V_b = 7^3$

$V_c = \dfrac{22}{7} \cdot r^2 \cdot 7$ $\qquad\qquad V_b = 343$

$V_c = 22r^2$

The volume of the cylinder is 31 cubic feet larger than the volume of the cube.

$$V_c = V_b + 31^*$$

$$22r^2 = 343 + 31$$

$$22r^2 = 374$$

$$r^2 = 17$$

$$r = \sqrt{17}$$

Answer: (D) $\sqrt{17}$

---

* Note: There is a psychological pull to add the 31 to the cylinder ($V_c + 31 = V_b$) because the cylinder is larger. However, we add the 31 to smaller object (the cube) to make it equal to the larger object ($V_c = V_b + 31$).

17. If we multiply $\dfrac{3b^2-4}{2}$ by 3, we get 282. What is the negative value of b?

(A) –10 (B) –7 (C) –6 (D) –8 (E) –9

Multiplying the expression $\dfrac{3b^2-4}{2}$ by 3 and setting the result equal to 282 gives

$$3\left(\dfrac{3b^2-4}{2}\right) = 282$$

$$\dfrac{9b^2 - 12}{2} = 282$$

$$9b^2 - 12 = 564$$

$$9b^2 = 576$$

$$b^2 = 64$$

$$b = \pm 8$$

The negative value of b is –8.

Answer: (D) –8

18. In a coffee harvest, 20 ounces of coffee are rejected out of every 90 pounds. What is the ratio of non-rejected coffee to rejected coffee? One pound = sixteen ounces.

(A) 19 : 3 (B) 56 : 1 (C) 26 : 1 (D) 71 : 1 (E) 32 : 3

Rejected coffee: 20 ounces

Non-rejected coffee: 90 pounds – 20 ounces = 90 pounds × 16 ounces – 20 ounces = 1420 ounces

$$\dfrac{\text{non-rejected coffee}}{\text{rejected coffee}} = \dfrac{1420}{20} = \dfrac{71}{1}$$

Answer: (D) 71 : 1

19. What is the range of the function $f(x) = 3 \cos 5x$?

(A) $-5 \leq y \leq 5$ (B) $-3 < y < 3$ (C) $-5 < y < 5$ (D) $-3 \leq y \leq 3$ (E) $-3.5 \leq y \leq 3.5$

(i) The constant preceding cos 5x tells us that the range (amplitude) extends from –3 to +3.
(ii) Alternatively, we can graph the function and actually see the range.

Answer: (D) $-3 \leq y \leq 3$

20. Carmen and Mandy run towards each other from towns A and B, respectively. The towns are 40 miles apart. Carmen runs at the rate of 8 mph. Mandy starts 1/2 hour later and runs 2 mph faster than Carmen. If they meet at 3:15 pm, at what time did Carmen start running?

(A) 12:45 pm   (B) 11:30 am   (C) 1:15 pm   (D) 1 pm   (E) 12:15 pm

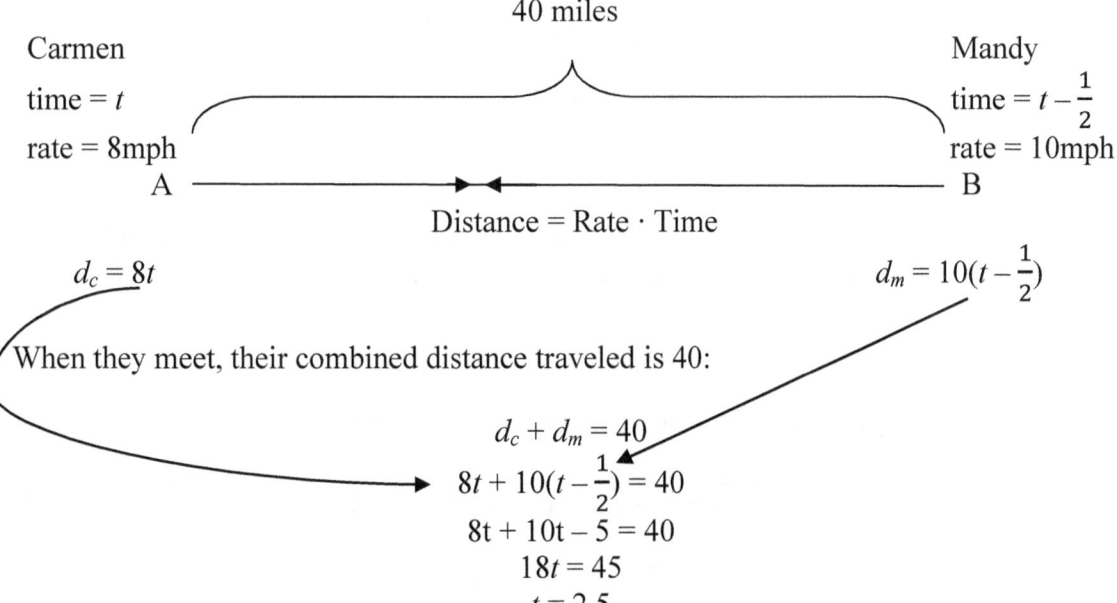

If they meet at 3:15 pm and it took them $2\frac{1}{2}$ hours to meet after Carmen began running, they started at 3:15 pm – 2.5 hours = 12:45 pm.

Answer: (A) 12:45 pm

21. A parabola is the set of points equidistant from a fixed line, called the directrix, and a fixed point, called the focus. Using this definition, develop an equation for a parabola with a focus of (2, 0) and a directrix $x = -2$, by setting the two distances equal.

(A) $y = \pm\sqrt{2x + 4}$ (B) $x = 6y^2$ (C) $y = \pm\sqrt{3x}$ (D) $y = 6x^2$ (E) $y = \pm\sqrt{8x}$

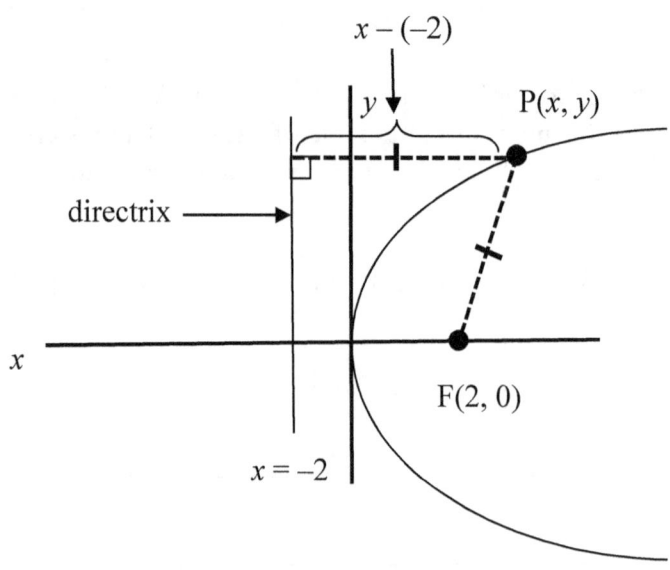

Since "a parabola is the set of points equidistant from a fixed line, called the directrix, and a fixed point, called the focus," we get from the figure:

$$PF = x - (-2)$$

Using the distance formula to calculate length of the segment PF gives

$$\sqrt{(x-2)^2 + (y-0)^2} = x - (-2)$$

$$\sqrt{(x-2)^2 + y^2} = x + 2$$

$$(x-2)^2 + y^2 = (x+2)^2$$

$$x^2 - 4x + 4 + y^2 = x^2 + 4x + 4$$

$$y^2 = 8x$$

$$y = \pm\sqrt{8x}$$

Answer: (E) $y = \pm\sqrt{8x}$

22. If the central angle of a circle measures 60° and the radius is 6, find the shaded area (the segment). Leave your answer in terms of π.

(A) $4\pi - 3\sqrt{2}$ (B) $6\pi - 8\sqrt{3}$ (C) $12\sqrt{3} - 4\pi$ (D) $6\sqrt{3} - 4\pi$ (E) $6\pi - 9\sqrt{3}$

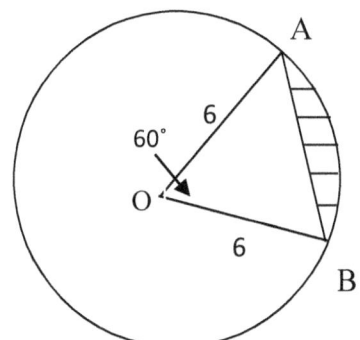
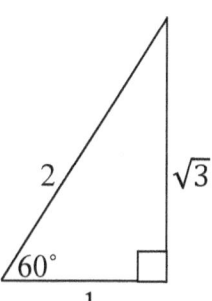

Area of sector AOB = $\frac{60}{360}$ × (area of the circle)

$= \frac{1}{6} \cdot \pi r^2$

$= \frac{1}{6} \cdot \pi \cdot 6^2$

Area of sector AOB = 6π

Area of triangle AOB = $\frac{1}{2} ab \sin C$
$= \frac{1}{2} \cdot 6 \cdot 6 \cdot \sin 60°$
$= 18\left(\frac{\sqrt{3}}{2}\right)$
$= 9\sqrt{3}$

Area of shaded region =

(Area of sector AOB) – (Area of triangle AOB) =

$6\pi - 9\sqrt{3}$

Answer: (E) $6\pi - 9\sqrt{3}$

23. The letters a, b and c represent 3 positive integers. If we double the value of a, reduce the value of b by $\frac{3}{5}b$ and multiply the value of c by $2\frac{1}{2}$, compare the value of the product of a, b and c with the product of the new values.

(A) 1 : 2 (B) 1 : 3 (C) 2 : 3 (D) 2 : 1 (E) 2 : 5

Original product: abc

New values of a, b, and c: $2a$, $b - \frac{3}{5}b$, $2\frac{1}{2}c$

New product: $(2a)\left(b - \frac{3}{5}b\right)\left(2\frac{1}{2}c\right) = 2abc$

Ratio of the original product to the new product:

$$\frac{Original}{New} = \frac{abc}{2abc} = \frac{1}{2}$$

Answer: (A) 1 : 2

24. A group of weightlifters competed. The results, by lifted weight, are listed below. Find the mean, correct to the nearest tenth.

(A) 149.1 (B) 157.4 (C) 158.6 (D) 166.5 (E) 161.2

Pounds	Number of Weightlifters
124	2
153	4
167	3
175	2
192	1

Forming the mean gives

$$\frac{2 \cdot 124 + 4 \cdot 153 + 3 \cdot 167 + 2 \cdot 175 + 1 \cdot 192}{12} =$$

$$\frac{1903}{12} \approx 158.6$$

Answer (C) 158.6

25. Let $f(x, y) = 3x^2 - 2y - 68 \geq 0$. If $y = 20$, what is the maximum negative value that x can assume?

(A) 4 (B) –5 (C) –6 (D) 7 (E) 8

$y = 20$:

$$3x^2 - 2y - 68 \geq 0$$

$$3x^2 - 2(20) - 68 \geq 0$$
$$3x^2 - 108 \geq 0$$
$$3x^2 \geq 108$$
$$x^2 \geq 36$$
$$\sqrt{(x^2)} = \sqrt{36}$$
$$|x| \geq 6$$
$$x \geq 6 \text{ or } x \leq -6$$

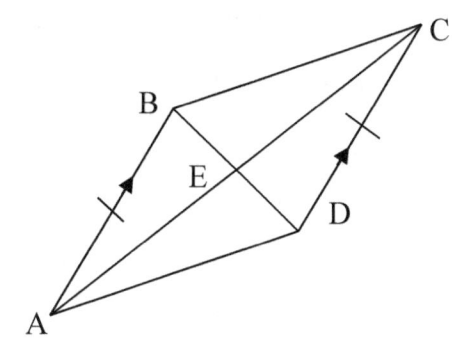

Answer: (C) –6   Largest possible negative value for x.

26. In the quadrilateral shown AB = CD and AB ∥ CD. List the maximum number of true statements.

(1) BC ≅ AD   (2) AB ≅ BC   (3) AE + BE = EC + ED   (4) AE ≅ BE   (5) BC ∥ AD

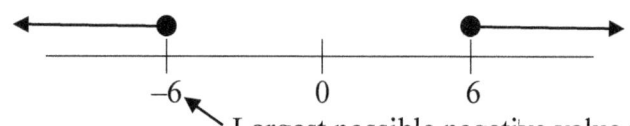

(A) 1, 2 (B) 2, 4 (C) 1, 2, 4 (D) 1, 3, 5 (E) 2, 4, 5

We've got a parallelogram, so opposite sides are congruent and parallel:

(1) BC ≅ AD
(5) BC ∥ AD
(3) The diagonals bisect each other, so AE + BE = EC + ED.

However, we do not know whether we have a square in which all sides are congruent.

Answer: (D) 1, 3, 5

27. In circle O, radius AO = 3 and $m\angle B = 60°$. Find the length of the minor arc AB. Leave your answer in terms of π.

(A) π (B) $\pi/3$ (C) $2\pi/3$ (D) 2π (E) $3\pi/4$

Triangle AOB is at least isosceles (two of its sides are radii of the circle), so the base angles are equal to 60° each. $m\angle AOB = 180° - 60° - 60° = 60°$.

$\dfrac{60}{360} = \dfrac{1}{6}$.

Circumference $= 2\pi r = 2\pi(3) = 6\pi$.

$m\stackrel{\frown}{AB} = \dfrac{1}{6} \cdot 6\pi = \pi$

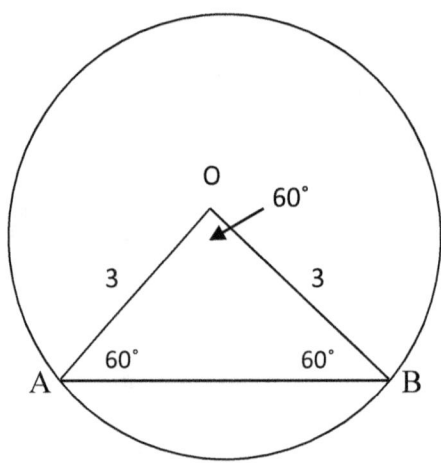

Answer: (A) π

28. Find the sum of the first 12 terms of the geometric sequence 2, 4, 8, 16,....

(A) $-2(1-2^{13})$ (B) $2(1-2^{13})$ (C) $-2(1-2^{13})$ (D) $-2(1-2^{12})$ (E) $-(1-2^{12})$

For the following formula, S_n = the sum, a = the first term, r = the common ratio and n = the number of terms:

$$S_n = \frac{a(1-r^n)}{1-r}$$

$$S_{12} = \frac{2(1-2^{12})}{1-2}$$

$$S_{12} = -2(1-2^{12})$$

Answer: (D) $-2(1-2^{12})$

29. A and B are two independent events out of many. If the probability of A occurring is .2 and the probability of B occurring is .6, what is the probability of A and not B?

(A) .12 (B) .24 (C) .08 (D) .32 (E) .48

$$P(A) = .2 \qquad P(\text{not } B) = 1 - .6 = .4$$

$$P(A) \text{ and } P(\text{not } B) = .2 \times .4 = .08.$$

Answer: (C) .08

30. Find the inverse of the logarithmic function $f(x) = \log_3(x + 2)$.

(A) $2^x - 3$ (B) $3^2 - x$ (C) $2^x + 3$ (D) $2 - 3^x$ (E) $3^x - 2$

$$f(x) = \log_3(x + 2)$$
$$y = \log_3(x + 2)$$

Switch x and y to form the inverse:

$$x = \log_3(y + 2)$$

Using the definition of a logarithm ($\log_b z = w$ iff $b^w = z$) yields

$$3^x = y + 2$$
$$y = 3^x - 2$$
$$f^{-1}(x) = 3^x - 2$$

Answer: (E) $3^x - 2$

31. The letters a, b and c are positive consecutive even integers and the letters d, e, and f are positive consecutive odd integers such that d is 3 more than twice b. Describe f in terms of a.

(A) $3a - 3$ (B) $2a + 6$ (C) $2a + 3$ (D) $2b + 7$ (E) $2a + 11$

$$a$$
$$b = a + 2$$
$$c = a + 4$$
$$d = 2b + 3 = 2(a + 2) + 3 = 2a + 7$$
$$e = (2a + 7) + 2 = 2a + 9$$
$$f = (2a + 9) + 2 = 2a + 11$$

Answer: (E) $2a + 11$

ACT Math Personal Tutor

32. The Fahrenheit temperature for boiling water is 212°. The Celsius temperature for boiling water is 100°. The Fahrenheit temperature for freezing water is 32°, while the corresponding Celsius temperature for freezing water is 0°. The formula linking the two scales is

$$F = \frac{9}{5}C + 32$$

At what temperature do the scales have the same value?

(A) 40°   (B) –40°   (C) 60°   (D) –60°   (E) –50°

When both scales are the same, F = C. Replacing F with C in the formula gives

$$C = \frac{9}{5}C + 32$$

Multiply by 5:

$$5C = 9C + 160$$

$$-4C = 160$$

$$C = -40$$

Answer: (B) –40

33. A train traveling at 80 mph enters a tunnel. If it takes 6 seconds from the time the train enters the tunnel until the first car reaches the far end of the tunnel, how long is the train? One mile = 5,280 feet.

(A) 648 ft.   (B) 586 ft.   (C) 680 ft.   (D) 704 ft.   (E) 726 ft.

There are 60 minutes in one hour = 60 seconds per minute × 60 minutes = 3,600 seconds in an hour.

So, 6 seconds is $\dfrac{6}{3,600} = \dfrac{1}{600}$ part of an hour.

At 80 mph, the train was traveling at 80 miles × 5,280 ft. = 422,400 ft. per hour.

Using the formula *Distance = Rate • Time*, in 6 seconds, the train traveled

$$\dfrac{1}{600} \times 422,400 \text{ ft.} = 704 \text{ ft.}$$

Answer: (D) 704 ft.

34. Find the radius of the circle shown if the area of the sector is 154/8 and the central angle is 45°. Let $\pi = 22/7$.

(A) 3   (B) 4   (C) 5   (D) 6   (E) 7

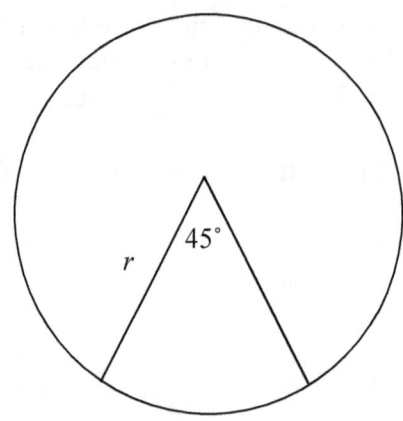

Area of the circle is πr^2.

The area of the sector is $\dfrac{45}{360} \cdot \pi r^2 = \dfrac{1}{8} \cdot \pi r^2$.

$$\dfrac{1}{8} \cdot \pi r^2 = \dfrac{154}{8}$$

Replace π with 22/7:

$$\dfrac{1}{8} \cdot \dfrac{22}{7} r^2 = \dfrac{154}{8}$$

Multiply by 7 and 8:

$$22r^2 = 1078$$

$$r^2 = 49$$

$$r = 7$$

Answer: (E) 7

35. Margaret is three times as old as her daughter. Four years ago Margaret was four times as old as her daughter was then. How old is Margaret now?

(A) 45 (B) 39 (C) 36 (D) 28 (E) 24

When assigning variables (say x), let the x represent the least known quantity. Here, the two unknown quantities are Margaret's age and her daughter's age. We know something about Margaret's age — namely, that it is three times her daughter's age. So, the least known quantity is the daughter's age:

	Age Now	Age 4 Years Ago
Daughter	x	$x - 4$
Margaret	$3x$	$3x - 4$

Four years ago Margaret was four times as old as her daughter was then:

$$3x - 4 = 4(x - 4)$$
$$3x - 4 = 4x - 16$$
$$x = 12$$
$$3x = 36$$

Answer: (C) 36

36. Which equation represents a move up by 3 units of the original equation $y = (x + 2)^2$?

(A) $y = x^2 + 3$ (B) $y = 3(x - 2)^2 + 2$ (C) $y = x^2 + 4x + 7$ (D) $y = x^2 + 2$
(E) $y = 2x^2 + 3$

The best way to solve this problem is to plug the given equations into your calculator.

Or we can reason it out.

$y = (x + 2)^2$ $y = x^2 + 4x + 7$

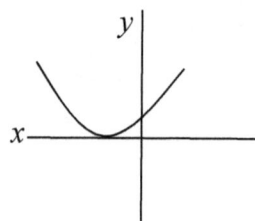
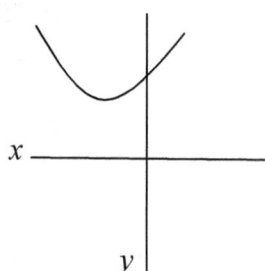

The original equation was $y = (x + 2)^2$.

To move up by 3, we'll have to add 3 to the original equation:

$$y = (x + 2)^2 + 3$$
$$y = (x^2 + 4x + 4) + 3$$
$$y = x^2 + 4x + 7$$

Answer: (C) $y = x^2 + 4x + 7$

37. If each interior angle of a regular polygon is 20° more than 3 times its adjacent exterior angle, how many sides does the polygon have?

(A) 12 (B) 8 (C) 6 (D) 10 (E) 9

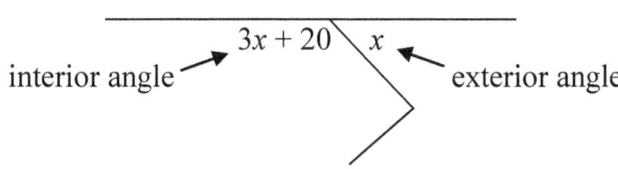

The two angles are supplementary (since they form a straight line):

$$(3x + 20) + x = 180$$
$$4x + 20 = 180$$
$$4x = 160$$
$$x = 40$$
$$3x + 20 = 3(40) + 20 = 140$$

The formula linking the number of degrees (a) in an individual interior angle of a regular polygon to the number of sides (n) in the figure is

$$a = \frac{(n-2) \cdot 180}{n}$$

$$140 = \frac{(n-2) \cdot 180}{n}$$

$$140n = 180n - 360$$

$$-40n = -360$$

$$n = 9$$

Answer: (E) 9

38. Jamal dives into a river and swims across at an angle of 45° with the terrain. When he touches land he has completed a swim of 2,400 feet. How wide is the river? Round your answer to the nearest foot.

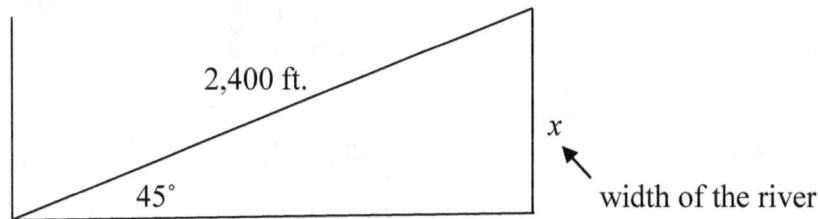

(A) 2,435 ft.   (B) 1,697 ft.   (C) 1,982 ft.   (D) 2,072 ft.   (E) 2,123 ft.

We'll use the sine function since it involves the opposite side (width of the river) over the hypotenuse (distance swam) of a triangle:

$$\sin 45° = \frac{x}{2,400}$$

$$\frac{1}{\sqrt{2}} = \frac{x}{2,400}$$

$$x = \frac{2,400}{\sqrt{2}} \approx 1,697$$

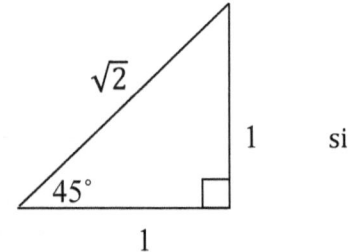 $\sin 45° = \dfrac{1}{\sqrt{2}}$

Answer: (B) 1,697 ft.

39. Jack jogs 12 miles at the rate of 8 mph. He walks back along the same path at the rate of 3 mph. What was his average speed?

(A) 5.2 mph   (B) 4.5 mph   (C) 4.6 mph   (D) 5.5 mph   (E) 4.4 mph

$$\xrightarrow{\underset{8 \text{ mph}}{12 \text{ miles}}} \qquad \frac{12 \text{ miles}}{8 \text{ mph}} = 1.5 \text{ hours}$$

$$\xleftarrow{\underset{3 \text{ mph}}{12 \text{ miles}}} \qquad \frac{12 \text{ miles}}{3 \text{ mph}} = 4 \text{ hours}$$

Solving the formula D = R • T (Distance = Rate • Time) for R gives $R = \frac{D}{T}$.

$$R = \frac{12 \text{ miles} + 12 \text{ miles}}{1.5 \text{ hours} + 4 \text{ hours}} = \frac{24 \text{ miles}}{5.5 \text{ hours}} \approx 4.36 \approx 4.4$$

Answer: (E) 4.4 mph

40. If $(x - y)(x - y) = 81$ and $y = -3$, find one value of x.

(A) 5   (B) –12   (C) –5   (D) 3   (E) –4

$$(x - y)(x - y) = 81$$
$$(x + 3)(x + 3) = 81$$
$$x^2 + 6x + 9 = 81$$
$$x^2 + 6x - 72 = 0$$
$$(x + 12)(x - 6) = 0$$
$$x = -12 \text{ and } x = 6$$

Answer: (B) –12

ACT Math Personal Tutor

41. Select the number line which expresses the range of values for x in the absolute inequality $|3x - 2| < 10$.

(A) ○———○ $-2\frac{2}{3}$ 0 4

(B) ●———○ $-2\frac{2}{3}$ 0 4

(C) ●———● $-2\frac{2}{3}$ 0 4

(D) ○———● $-2\frac{2}{3}$ 0 4

(E) ○   ● $-2\frac{2}{3}$ 0 4

$$|3x - 2| < 10$$
$$-10 < 3x - 2 < 10$$
$$-8 < 3x < 12$$
$$-2\frac{2}{3} < x < 4$$

Answer: (A)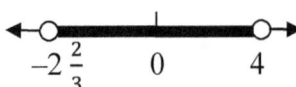

42. Find the reflection of point P (4, 2) across the line $y = x$.

(A) (2, 4)   (B) (4, 3)   (C) (3, 5)   (D) (–3, 4)   (E) (4, –3)

Merely, reverse the coordinates:

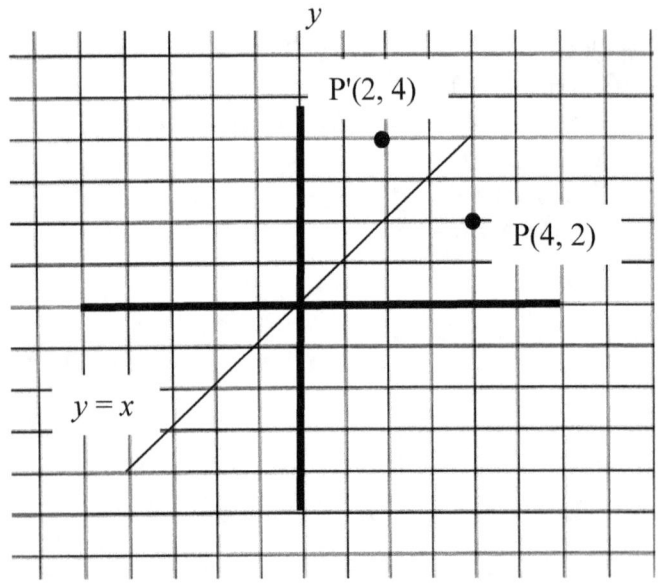

Answer: (A) (2, 4)

43. Find the *y*-coordinate of point A on the ellipse shown, with center at the origin. Leave your answer in radical form.

(A) $\sqrt{5.25}$ (B) $\sqrt{6.75}$ (C) $\sqrt{3.85}$ (D) $\sqrt{4.15}$ (E) $2\sqrt{4.85}$

The standard from for the equation of an ellipse with center at the origin is

$$\frac{x^2}{a^2}+\frac{y^2}{b^2}=1$$

where *a* is 1/2 the length of the horizontal axis, *b* is 1/2 the length of the vertical axis, and *x* and *y* are coordinates of any point on the ellipse.

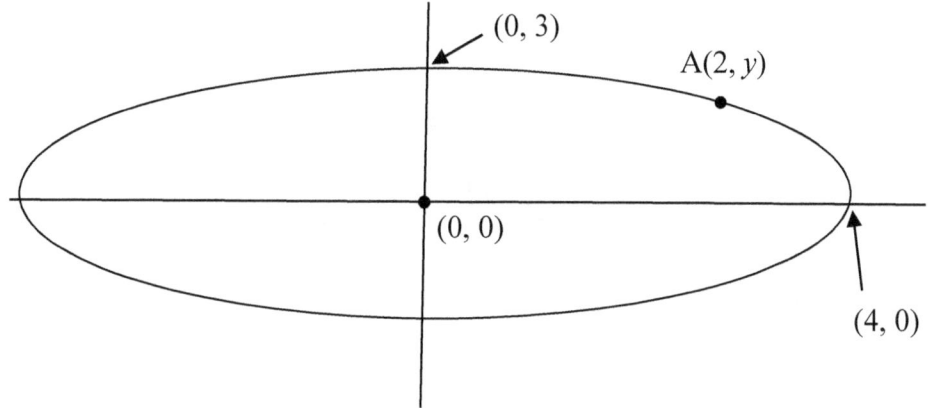

From the figure, we see that the *x*-coordinate of point A is 2, that ½ the length of the horizontal axis, *a*, is 4, and that ½ the length of the vertical axis, *b*, is 3:

$$\frac{2^2}{4^2}+\frac{y^2}{3^2}=1$$

$$\frac{4}{16}+\frac{y^2}{9}=1$$

Multiply by the LCD 144:

$$36 + 16y^2 = 144$$

$$16y^2 = 108$$

$$y^2 = 6.75$$

$$y = \sqrt{6.75}$$

Answer: (B) $\sqrt{6.75}$

44. If $\dfrac{3}{5}$ of 3% of $b = 81$, find $\dfrac{2b}{3}$.

(A) 6,000   (B) 4,500   (C) 6,500   (D) 3,000   (E) 2,500

Translating the statement "$\dfrac{3}{5}$ of 3% of $b = 81$" into an equation yields

$$\dfrac{3}{5} \times .03b = 81$$

$$\dfrac{.09}{5}b = 81$$

$$.09b = 405$$

$$b = 4{,}500$$

$$\dfrac{2b}{3} = \dfrac{2 \cdot 4{,}500}{3} = 2 \cdot 1{,}500 = 3000$$

Answer: (D) 3,000

Test 5

45. Solve for the two values of x in the equation $\sqrt{10x+1} = \sqrt{9x-8} + 1$.

(A) $x = 5, 6$ (B) $x = 7, 4$ (C) $x = 3, 8$ (D) $x = 9, 11$ (E) $x = 8, 12$

Square both sides of the equation:

$$\left(\sqrt{10x+1}\right)^2 = \left(\sqrt{9x-8}+1\right)^2$$

$$10x + 1 = 9x - 8 + 2\sqrt{9x-8} + 1$$

$$x + 8 = 2\sqrt{9x-8}$$

$$(x+8)^2 = \left(2\sqrt{9x-8}\right)^2$$

$$x^2 + 16x + 64 = 4(9x - 8)$$

$$x^2 + 16x + 64 = 36x - 32$$

$$x^2 - 20x + 96 = 0$$

$$(x-8)(x-12) = 0$$

$$x = 8 \text{ or } x = 12$$

Answer: (E) $x = 8, 12$

46. Name the point in the first quadrant where the graphs of the functions $f(x) = 6/x$ and $g(x) = x$ cross?

(A) $(\sqrt{3}, \sqrt{3})$ (B) $(2\sqrt{3}, -2\sqrt{3})$ (C) $(\sqrt{6}, \sqrt{6})$ (D) $(\sqrt{5}, -\sqrt{5})$ (E) $(3\sqrt{3}, -3\sqrt{3})$

Solve the problem graphically:

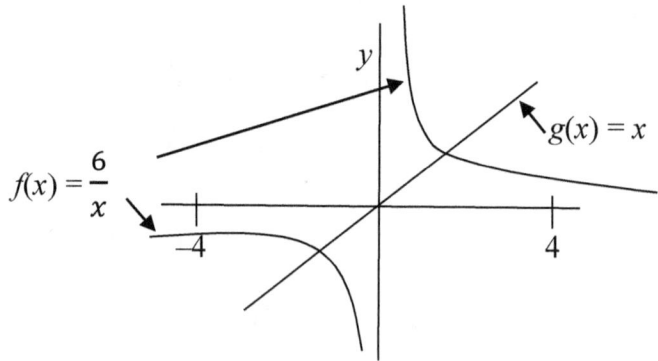

Solve the problem algebraically:

When the graphs cross, their y-values will be equal:

$$f(x) = g(x)$$
$$\frac{6}{x} = x$$
$$6 = x^2$$
$$x = \pm\sqrt{6}$$

$x_1 = +\sqrt{6}$ $\qquad$ $x_2 = -\sqrt{6}$

$y_1 = \dfrac{6}{x_1} = \dfrac{6}{\sqrt{6}} = +\sqrt{6}$ $\qquad$ $y_2 = \dfrac{6}{x_2} = \dfrac{6}{-\sqrt{6}} = -\sqrt{6}$

$(x_1, y_1) = (\sqrt{6}, \sqrt{6})$ $\qquad$ $(x_2, y_2) = (-\sqrt{6}, -\sqrt{6})$

The two graphs cross at $(\sqrt{6}, \sqrt{6})$ and at $(-\sqrt{6}, -\sqrt{6})$.

The graphs cross at two points.

Answer: (C) $(\sqrt{6}, \sqrt{6})$

47. Five pounds of Colombian coffee at $4.90 per pound are mixed with eight pounds of Venezuelan coffee to make a mixture at $5.70 per pound. What is the price of the Venezuelan coffee?

(A) $6.20   (B) $5.40   (C) $5.10   (D) $5.30   (E) $6.30

Let x = the price of the Venezuelan coffee.

$$5(4.90) + 8x = 13(5.70)$$
$$24.50 + 8x = 74.10$$
$$8x = 49.60$$
$$x = 6.20$$

Answer: (A) $6.20

48. In right triangle ABC, DE ∥ AB, AC = 10, AB = 15, DE is 9 more than BE, and DC = 8. Find BE.

(A) 3   (B) 5   (C) 9   (D) 12   (E) 8

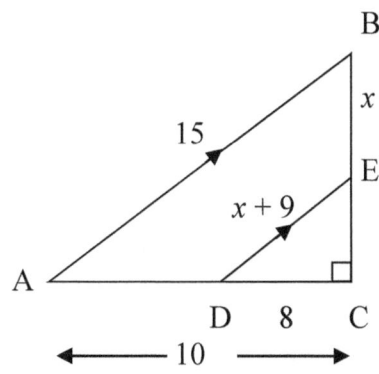

We have similar triangles, △ABC and △DEC, so the corresponding sides are in proportion:

$$\frac{x+9}{15} = \frac{8}{10}$$

$$10x + 90 = 120$$

$$10x = 30$$

$$x = 3$$

Answer: (A) 3

49. What is the width of the parabola representing the equation $y^2 = 12x$ at $x = 3$?

(A) 8 (B) 6 (C) 10 (D) 12 (E) 14

A parabola representing an equation of the form $y^2 = 4px$ faces sideways.

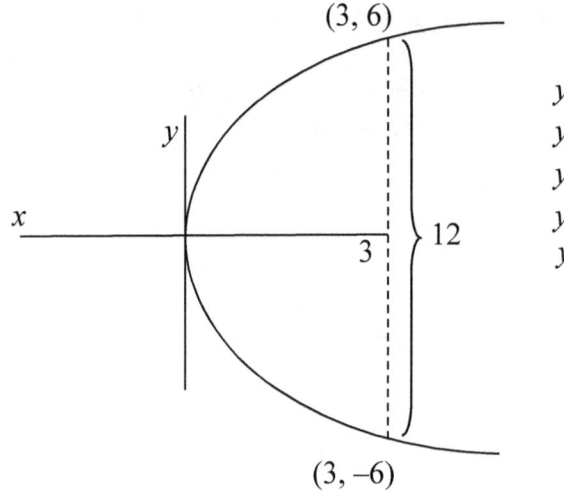

$y^2 = 12x$
$y = \pm\sqrt{12x}$
$y = \pm\sqrt{12 \cdot 3}$
$y = \pm\sqrt{36}$
$y = \pm 6$

Answer: (D) 12

50. If $\dfrac{1}{r} = \dfrac{1}{s} - \dfrac{1}{t}$, find the value of r in terms of s and t.

(A) $\dfrac{s+t}{t-s}$ (B) $\dfrac{st}{t-s}$ (C) $\dfrac{t-s}{t+s}$ (D) $\dfrac{s+t}{s-t}$ (E) $\dfrac{ts}{s+t}$

Multiply by the LCD rst to clear the fractions:

$$st = rt - rs$$

$$st = r(t - s)$$

$$r = \dfrac{st}{t-s}$$

Answer: (B) $\dfrac{st}{t-s}$

51. We have 8 different colored paints. We'll call them *a, b, c, d, e, f, g* and *h*. We want to paint 3 rooms. However, we don't want to use paint *b* in rooms 1 or 2. Once a paint is used, we cannot use it again in another room. In how many different ways can we paint the three rooms, using only one color per room?

(A) 336 (B) 384 (C) 480 (D) 252 (E) 216

We can't use *b* in room 1, so the only available colors are *a, c, d, e, f, g* or *h*, 7 colors. For room 2 we drop one of *a, c, d, e, f, g* or *h*, say *a*, so we have 6 colors available. For room 3 we drop one color from *c, d, e, f, g* or *h*, say *c*, but we can use *b*, so we have 6 colors available. For illustrative sake, we're using these letters.

a, c, d, e, f, g, h		*c, d, e, f, g, h*		*b, d, e, f, g, h*	
7	×	6	×	6	= 252
room 1		room 2		room 3	

Answer: (D) 252

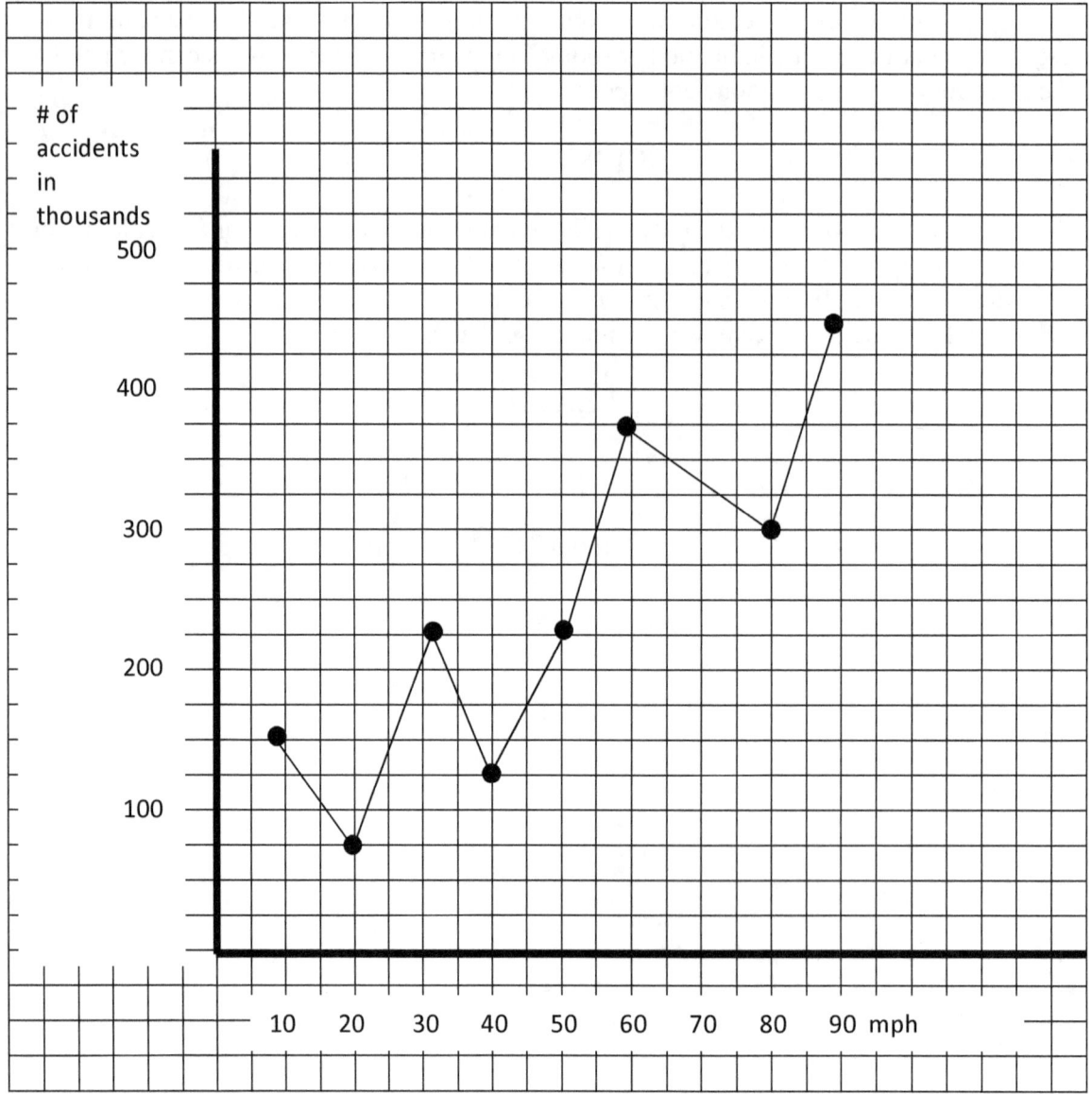

52. What was the percent increase in the number of accidents in driving at 40 mph compared to 80 mph? Round off to the nearest percent.

(A) 90% (B) 120% (C) 140% (D) 180% (E) 330%

At 40 mph there were 125,000 accidents. At 80 mph there were 300,000 accidents.

$$Percent\ Increase = \frac{Change}{Original\ Amount} = \frac{300,000 - 125,000}{125,000} = \frac{175,000}{125,000} = 1.4 = 140\%$$

Answer: (C) 140%

53. If the volume of this triangular prism is 96 and the base of the triangle is 6 and the height of the triangle is 4, find the length of the prism.

(A) 6 (B) 10 (C) 8 (D) 12 (E) 9

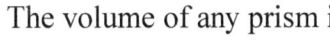

The volume of any prism is

(Area of base) • (Length)

(Area of triangle) • (Length)

$(\frac{1}{2}bh) \cdot (l)$

$\frac{1}{2}bhl$

$V = 96, b = 6, h = 4$:

$96 = \frac{1}{2}(6)(4)l$

$96 = 12l$

$l = 8$

Answer: (C) 8

54. Find the value of y in the trigonometric equation $y = 4 \cos x + 1$ when $x = 2\pi$.

(A) 5 (B) 2 (C) 3 (D) 9 (E) 8

$y = 4 \cos 2\pi + 1$
$= 4 \cdot 1 + 1$ since $\cos 2\pi = 1$
$= 5$

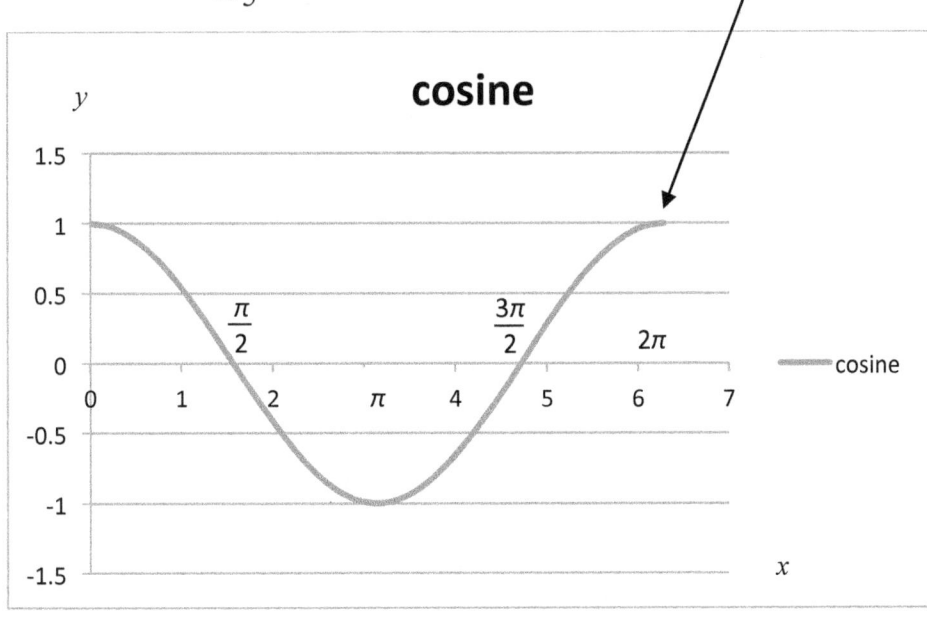

Answer: (A) 5

55. The center of a circle is located on the xy-plane at (−1, 2). Its radius is $\sqrt{29}$. At which positive point does the circle cross the x-axis?

(A) (3, 0)   (B) (0, 3)   (C) (4, 0)   (D) (5, 0)   (E) (0, 4)

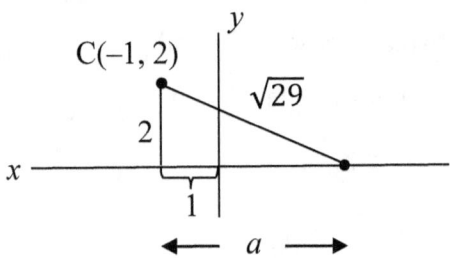

Use the Pythagorean Theorem to find side *a*.

$$a^2 + 2^2 = \left(\sqrt{29}\right)^2$$
$$a^2 + 4 = 29$$
$$a^2 = 25$$
$$a = 5$$

$$5 - 1 = 4$$

So, the positive point where the circle crosses the x-axis is (4, 0).

Answer: (C) (4, 0)

56. Find the area of the triangle enclosed by the three squares illustrated below. Leave your answer in radical form.

(A) 20   (B) 8   (C) 18   (D) 12   (E) 6

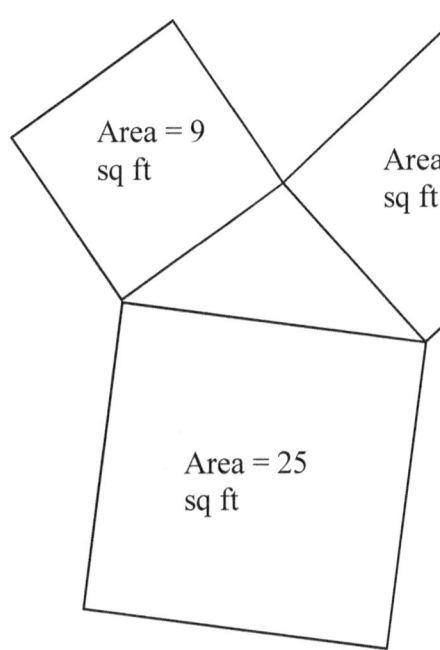

The squares have sides 3, 4 and 5, and the enclosed △ has those sides.

Note that sides 3, 4 and 5 satisfy the Pythagorean Theorem:

$$3^2 + 4^2 = 5^2$$
$$9 + 16 = 25$$
$$25 = 25$$

Hence, we have a right triangle with right angle formed by the sides of lengths 3 and 4. Let's take 3 to be the base and 4 to be the height of the triangle. Therefore, the area is

$$A = \frac{1}{2}bh = \frac{1}{2}(3)(4) = 6$$

Answer: (E) 6

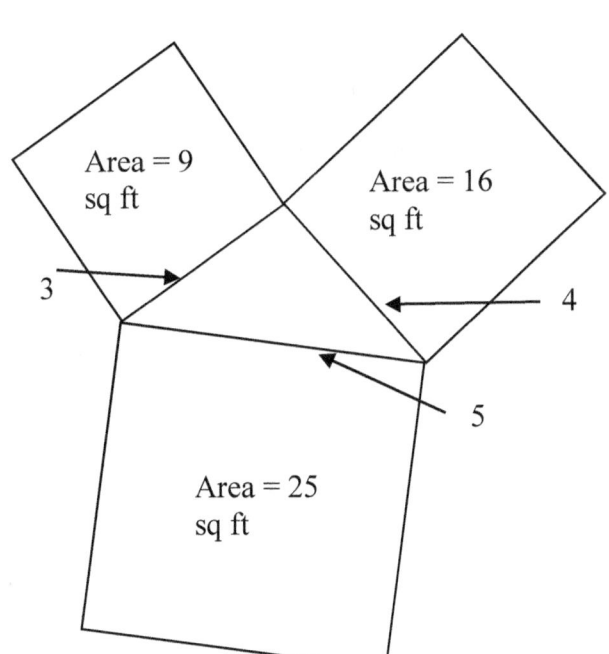

57. There are 4 green balls, 6 red balls and 3 yellow balls in a container. What is the probability of first selecting a red ball, not replacing it, and then selecting a yellow ball?

(A) 7/19 (B) 3/26 (C) 8/27 (D) 5/16 (E) 4/9

Probability of first selecting a red ball: 6/13

Probability of selecting a yellow ball if the red ball is not replaced: 3/12

Probability of both events occurring:

$$\frac{6}{13} \cdot \frac{3}{12} = \frac{18}{156} = \frac{3}{26}$$

Answer: (B) 3/26

58. If one point on the graph of the function $y = (x - 2)^3$ is $A(4, y_1)$, what is the equation of the straight line parallel to the x-axis which passes through that point?

(A) $y = 8$ (B) $x = 8$ (C) $y = 4x$ (D) $x = 6y$ (E) $y = -4$

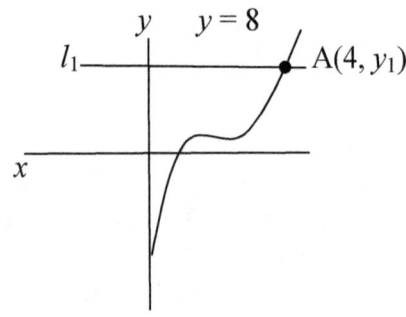

$y = (x - 2)^3$

$x = 4: y = (4 - 2)^3$

$y = 8$

The equation of the horizontal straight line passing through $(4, 8)$ is $y = 8$.

Answer: (A) $y = 8$

59. Which of the lines below has a slope of –2?

(A)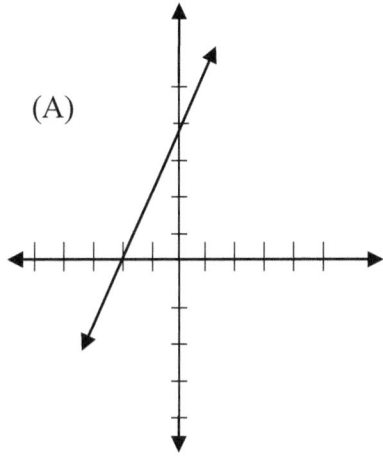

(B)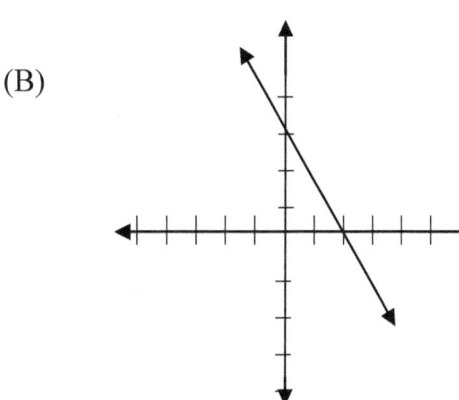

(C)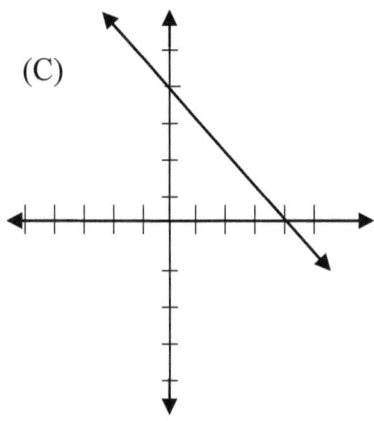

(D)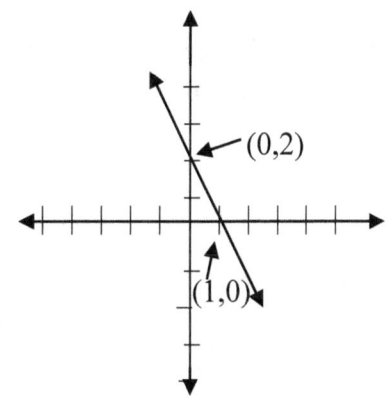

Slope $= \dfrac{\Delta y}{\Delta x} = \dfrac{y_2 - y_1}{x_2 - x_1} = \dfrac{2-0}{0-1} = \dfrac{2}{-1} = -2$

Answer: (D)

(E)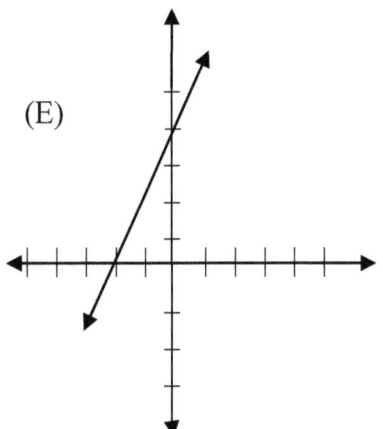

60. Find the average rate of change of the function displayed from $x = -4$ to $x = 3$.

(A) 2/5 (B) 5/8 (C) 3/7 (D) 4/9 (E) 5/7

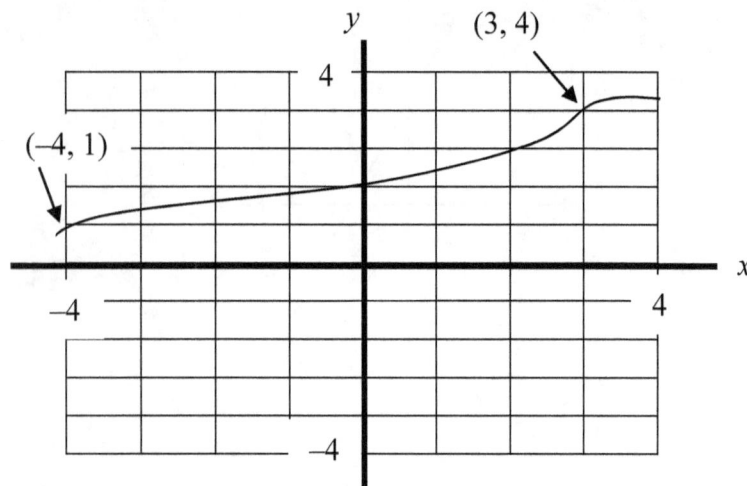

The average rate of change of a function between two points is the slope of the line connecting those two points, so

$$\text{Slope} = \frac{\Delta y}{\Delta x} = \frac{y_2 - y_1}{x_2 - x_1} = \frac{4-1}{3-(-4)} = \frac{3}{7}$$

Answer: (C) 3/7

Answer Key to Test 5

1. D
2. A
3. B
4. A
5. B
6. B
7. C
8. D
9. E
10. A
11. E
12. A
13. C
14. A
15. B
16. D
17. D
18. D
19. D
20. A
21. E
22. E
23. A
24. C
25. C
26. D
27. A
28. D
29. C
30. E
31. E
32. B
33. D
34. E
35. C
36. C
37. E
38. B
39. E
40. B
41. A
42. A
43. B
44. D
45. E
46. C
47. A
48. A
49. D
50. B
51. D
52. C
53. C
54. A
55. C
56. E
57. B
58. A
59. D
60. C

ACT Math Personal Tutor

Converting Raw Scores to Scaled Scores

The raw score results of each subject area of the ACT are converted to scaled scores ranging from 36 down to 1. The scaled scores of the four subject areas — English, Mathematics, Reading and Science — are then added together and averaged in order to get a composite score.

Converting Raw Math Scores to Scaled Scores

Raw Score	Scaled Score	Raw Score	Scaled Score
60	36	27–29	19
57–58	35	25–26	18
56	34	22–24	17
55	33	18–21	16
54	32	15–17	15
52–53	31	12–14	14
50–51	30	09–11	13
48–49	29	08	12
46–47	28	07	11
43–45	27	06	10
41–42	26	05	9
39–40	25	04	8
37–38	24	03	7
35–36	23	02	5
33–34	22	01	3
31–32	21	–	–
30	20	0	1

Converting Scaled Math Scores to Percentiles

Scaled Score	Percentile	Scaled Score	Percentile
36	99	18	43
35	99	17	37
34	99	16	27
33	98	15	15
32	97	14	6
31	96	13	2
30	95	12	1
29	93	11	1
28	91	10	1
27	88	9	1
26	84	8	1
25	78	7	1
24	73	6	1
23	67	5	1
22	62	4	1
21	57	3	1
20	53	2	1
19	49	1	1

www.ingramcontent.com/pod-product-compliance
Lightning Source LLC
Chambersburg PA
CBHW082108230426
43671CB00015B/2636